INTELLIGENT DESIGNER:

EVOLUTION FOR POLITICIANS

John Janovy, Jr.

Intelligent Designer: Evolution for Politicians, First Edition. Copyright © 2011 by John Janovy, Jr. All rights reserved. Printed in the United States of America. No part of this book may be used or reproduced in any manner whatsoever without written permission except in the case of brief quotations (600 words or fewer) in scholarly publications, articles, or reviews. For information contact the author at jjparasite@hotmail.com.

Chapters 6, 16, 18, and 20 appeared in the author's previous book, *Pieces of the Plains: Memories and Predictions from the Heart of America* (J&L Lee, Lincoln, NE).

Designed by John Janovy, Jr.

ISBN-13: 978-146638786
ISBN-10: 1466387386

Books by John Janovy, Jr.
Keith County Journal
Yellowlegs
Back in Keith County
On Becoming a Biologist
Fields of Friendly Strife
Vermilion Sea
Dunwoody Pond
Comes the Millennium (as Jack Blake)
Foundations of Parasitology (with Larry Roberts)
Biodiversity: A Primer (with Amanda Snyder)
Ten Minute Ecologist
Teaching in Eden
Outwitting College Professors
The Ginkgo
Conversations between God and Satan
Pieces of the Plains
Tuskers
Dinkle: A Spiritual Biography
Intelligent Designer

Table of Contents:

Foreword – 7
1. What is Intelligence? – 11
2. What is Design? – 23
3. What is Intelligent Design? – 35
4. What is Complexity? – 47
5. What is Creationism? – 59
6. What is Science? – 73
7. What is Religion? – 85
8. What is a Conflict between Science and Religion? – 95
9. Why are Science and Religion in Conflict? – 107
10. What is Evolution? – 119
11. What Kinds of Organisms Share the Planet with Us? – 135
12. What is Taught in Biology Class? – 145
13. What Should be Taught in Biology Class? – 157
14. What is the Meaning of "Scientific Literacy"? – 171
15. Why is Scientific Literacy of Such Importance? – 185
16. Why are Politicians so Scientifically Illiterate? – 197
17. Is "Evolution" Dangerous? – 209
18. What is a Human Being? – 219
19. Are Humans Evolving? – 233
20. What will Human Life be Like in a Couple of Thousand Years? – 245

References and Sources – 259
About the Author – 267

Foreword

I wrote this book because I'm biology professor who has a lot of non-biology professor friends and a lot of students with strong religious beliefs but little if any scientific education. After many heated conversations with these friends, and equally heated e-mail exchanges with students, always asking myself exactly what was going on in their minds, I came to the conclusion that they, like many if not most of my fellow Americans, were simply able to compress a massive amount of history and complex human interactions into single words—think "church," "God," "immigration," "abortion," "homosexuality," or "evolution"—then act on the compressed version. The last four of these words, of course, represent rather obvious biological phenomena, about which a biology prof ought to have something to say.

Often, if not regularly, my friends' actions, based on their highly simplified view of complex phenomena, involve voting for, and giving money to, candidates for public office. These candidates often are more ignorant about a whole lot of subjects than are my friends and students. If such candidates were running for local weed control board, then their ignorance wouldn't seem to pose much of a problem for society. But when they're running for school board, state legislature, or American President, then their scientific illiteracy is a real public health hazard.

Lewis Thomas (*Lives of a Cell*), one of our most literate medical writers, likened the entire planet to a cell and made a fairly interesting philosophical case for doing so. Thomas led me to wonder about nations, and whether the cell metaphor might also apply to them. Thus I began to study the United States in terms of its components (metaphorical organelles such as my friends, especially their occupations), interdependency between those components (commerce as metabolism), and most importantly, the environment occupied by this cell-nation, that is, its culture medium. As a result of this study, I came to the conclusion we were a sick cell-nation, mainly because of our environment. This environment, it turns out, is not in the form of a protein soup. Instead, it's an idea soup, a concept soup, a soup of words, emotions, impresssions, and beliefs. And the words that seem to be most powerful, most capable of generating emotions and challenging beliefs, are those four biological hot buttons: immigration, abortion, homosexuality, and evolution.

The strongest of our beliefs, namely, the religious ones, seem to dictate our political actions to a remarkable degree, given that we are a nation in which church and state are supposed to be legally separated. Often such influence is quite counter to the national good, especially when wielded by willfully ignorant people promoting scientific illiteracy in the name of religion, especially conservative Christian religion. I'm speaking, of course, about evolution and creationism. In our Third Millennium America, any candidate for public office must appear completely clean on matters of American ideology. Nobody who runs on a secular humanist platform is likely to get past the primaries. We're for the troops, jobs, church, and family, and against sex, taxes, immigration, abortion, gay rights, and evolution. For the word "evolution," however, you can substitute "science literacy." When you're against science literacy, then you're a germ infecting our cell-nation.

Candidates for public office actually are candidates for public power, power that will be exercised in ways that affect us all. A woman's right to obtain a safe, clean, abortion is an issue that will never be resolved, simply because of our emotional response to the image of a baby, regardless of our nation's miserable record of child abuse, foster care, health care, and juvenile detention. Gay rights are an issue that will be resolved by demographic change. Old conservative religious white men in positions of public power eventually will die off and be replaced by people who have knowingly lived among and worked with homosexuals and thus are not convinced sexual orientation is much of a sin, and certainly not enough of a sin to be denied equal opportunity under the law. When the Iowa Supreme Court upholds gay marriage, you know that our society is outgrowing one of its most pervasive prejudices.

But there is little chance that as a nation we will solve the immigration problem in the foreseeable future; we are simply too fearful of "the other," that is, anyone who is brown, not already a citizen, and whose primary language is not American English. Evolution, on the other hand, is a relatively strange case because in the public mind, even in the absence of even a shred of evidence to support this position, the word itself equates to all sorts of evilness ranging from godlessness to National Socialism.

Politically active creationists have managed to keep the word "evolution" in the center of the debate over human origins, which to be honest is an entirely political one, that is, a culture war, but one with very real consequences for our nation. Scientists have established, unequivocally, that humans evolved from non-human primate ancestors. We may not know *exactly which one* of these ancestral species was our progenitor, but based on the scientific evidence, it is abundantly clear that there now are several good candidates among the extinct primates that inhabited Africa five to ten million years ago. Those discoveries have not done much to erode the belief, among conservative Christians and those who are scientifically illiterate, that we the people are not "descended from monkeys" but instead are made by God in God's image.

Belief may trump knowledge in the political arena, but never in the arena of life on Earth. We will, eventually, live out our species' destiny as really smart apes, whether that destiny be obliteration by nuclear weapons of mass destructtion, starvation and genetic bottleneck once our population reaches its so-called stable limits, or destruction of the biosphere to the point that our populations are no longer sustainable. The time when we get our biological come-uppance cannot be predicted accurately, but it can certainly be estimated, based on per-capita water use and the global petroleum supply. Most scientists estimate that time somewhere near the end of the current century, that is, 2060-2100.

Scientists also estimate that the population will level off about that same time. That estimate actually is a testable hypothesis, an assertion, although not necessarily one that can be considered an "experiment" as we usually define the term. Nevertheless, humans, including most American children born from 1990 to 2010, those currently old enough to vote, or who will become old enough to vote before many current politicians are up for re-election, will get to see the result of humanity's test of the scientific hypothesis that exponential growth can be sustained forever, especially when sustained by fixed and diminishing resources. The fact is that we already know the result (it cannot). The general principle that *exponential growth cannot be sustained indefinitely against fixed resources* is so well known that our collective vision of the future ought to be a no-brainer.

Unfortunately, for the species and our nation, our leaders evidently do not understand this principle, and certainly are not equipped to explain it to an electorate that believes otherwise. We are a nation that depends, like life's blood itself, on science literacy. That's why I wrote this book. It's one person's attempt to solve the problem.

John Janovy, Jr.
October, 2011

1. What is Intelligence?

> . . . *the length of childhood, adolescence, and life expectancy for a chimpanzee are double those for a macaque, whose brain is about one fourth the size of the chimpanzee's. Likewise, our life cycle is much longer than that of a chimpanzee.*
> —Juan Luis Arsuaga (*The Neanderthal's Necklace*)

The most basic definition of "intelligence" is the capacity to learn, primarily from experience. By "learn" I mean to acquire information, retain it, and subsequently use it in some way. For smart animals such as dogs and horses, information comes in a variety of forms: sights, sounds, actions of other animals, and most importantly, smell. For very smart animals such as chimpanzees, information comes not only in those same forms, but also from parents, siblings, and peers, and routinely involves behavior that ends up being an example of how to be a successful chimp. For extremely smart animals like humans, information comes in a truly bewildering, in fact almost indescribable, variety of forms: books, magazines, television, e-mail, text messages, all in addition to sights, sounds, and smells emanating from nature and our constructed environments such as large cities and small villages.

Thus if you arrange animals in some *scala naturae*, with worms at the bottom and humans at the top, you also discover that the variety of information accessible to animals increases along this scale, bottom to top, but nevertheless takes a quantum leap in both quantity and qualities between chimps and people. Most scientists, of course, reject the *scala naturae* as a valid representation of anything other than our own opinion of ourselves. Plants do not have what we would call "intelligence" regardless of the fact that they respond to their environments, including the presence of other plants, albeit slowly, and through differential growth rather than movement powered by muscles. Plants do use information about temperature, moisture, light, and chemical compounds, but to a human, especially a non-scientist, such use is fairly passive and subtle. There's nothing subtle, however, about the information use gap between our closest non-human relatives the chimps, and ourselves; this gap is simply enormous.

The retention and use of information are key characteristics of intelligence as we typically use the term. If you're a squirrel and you can't remember where you buried an acorn, you may be in trouble come winter. On the other hand, if you can smell acorns in the ground, especially through the snow, then you don't have to remember so much. So squirrels are a good example of one important principle about intelligence, namely, that if you have ready and virtually guaranteed access to information, such access relieves you of the responsibility for retaining this information.

This principle is an extremely important one that tends to shape our lives in the 21st Century, the so-called "information age." If you can find out anything you want about any subject just by clicking on Google®, then you don't have to remember very much. On the other hand, if the Internet becomes a substitute for real, some might say "classical," learning, retention, and assimilation, then we have become merged with our machines at least to some degree.

When we humans use the word "intelligent," we imply a number of things beyond simple information retention and use, for example, ability to solve new problems, creation of technology, analytical skills, insight, use of language, and behavior perceived to be characteristic of "smart" people (reading difficult books with no pictures, listening to classical music, appreciating abstract expressionist art, etc.) So even among members of our own species, we recognize not only different levels, but also different kinds of intelligence.

Many of us know engineers, attorneys, and physicians, and most of us would consider these kinds of professionals to be intelligent. We also know skilled workers or craftsmen who are very good at what they do, and often we have deep respect for such people. The gentlemen who perform our annual furnace and air conditioner maintenance, and who fix our automobiles, come immediately to mind. We would never consider these people "dumb," and if we did, we probably would not let them replace an air conditioner or furnace, especially after watching one of them actually do it single-handedly.

In the visual and performing arts, we rarely refer to intelligence but instead talk, and write, about "talent." We also know that some people are much better artists and musicians than others, and we also know that at least some of this variation is inborn. Such gifted artists can do things that most other people cannot and like engineers, artists also practice and do formal exercises to develop and maintain their skills as well as explore new ways of artistic expression. Furthermore, artists remember particularly effective training activities, apply this acquired knowledge to modify their own performances, and if they are in a teaching situation, use their knowledge to improve the skills and work of their students. We still call these artists "talented," or in the case of some of them "visionary," instead of "smart," but in fact they are doing what smart people do, just in disciplines not always associated with "intelligence" as we normally use the term.

Various human capacities that we interpret as "smart" or requiring "intelligence" can be seen with even a cursory examination of our local communities. In fact, we are so diverse in our abilities, talents, and interests that it is almost impossible to define "intelligence" in human terms, except perhaps in an historical context. As a species we have built nuclear weapons and power plants, sent machines on exploratory missions to other planets, invented mathematics and physics then used those inventions to infer the fundamental nature of the universe, written music and made art that communicates in some mysterious way across time, space, and cultures, figured out numerous species' genomes (including our own), constructed magnificent cities, described a million species of plants, animals, and microbes, invented electronic communications, and cured the world of smallpox. Regardless of what you might think about your elected officials or a football coach who called the wrong play on third and short, history shows us that *Homo sapiens* is an *extremely* intelligent, and highly capable, species.

In the monotheistic religions, not only is God omnipotent—all powerful—He also is omniscient; that is, He knows everything. So God must be the most intelligent of all entities known, although to be brutally honest, there is plenty of evidence, including many passages from the Old Testament, that *knowing* everything doesn't mean that God behaves much differently than many humans, especially those in positions of great power, for example, kings and military commanders, but kings in particular. In fact, there are passages in the Bible that seem to reveal a rather capricious behavior on the part of God, a good example being Genesis 3 where God decides that Adam and Eve are somehow better off not knowing the difference between good and evil. In my humble opinion, if there is anything that humanity needs today, it's a clear sense of the difference between good and evil, as well as the courage to do good instead of evil.

"God's will" and "acts of God" are phrases that disguise a seeming capriciousness, for example, when a misfortune happens to some people and not to others, or when contracts made by humans are voided by natural disasters. The fact that we may have no rational explanation for why such events happen to some people and not others just deepens the mystery behind the omniscient and omnipotent supernatural being. Nevertheless, it seems fairly stupid and ignorant to consider so-called legal "acts of God," for example, Hurricane Katrina or the January, 2010 earthquake in Haiti, real acts of an omniscient and omnipotent God, although some people obviously believe such events are indeed punishment for those who deserve the misfortune. But regardless of what the wacko Wesboro Baptist Church tells us about God's vengeance, every half-way educated human knows full well that hurricanes, earthquakes, and tornadoes are fairly common acts of Earth, not of God.

So it's arguable whether knowledge is the same as intelligence, especially if the latter trait is defined as actions that are driven not only by knowledge, but also by insight, memory, experience, and the ability recognize and solve new problems. In other words, if the intelligent design proponents are correct, then God was plenty smart enough to build a universe, then build a cell on one of its planets, but not smart enough to build human beings who can get along with one another peacefully. There could easily be many versions of this analysis of God's intelligence. We could also say He was intelligent enough to build a human being but not smart enough to make a human that would use his or her own intelligence to solve major environmental and social problems so that other humans could live dignified, safe, and enjoyable lives. Among the many versions of God's intelligence would be one that claims He is indeed omniscient and omnipotent, but simply chose to build a human species with flaws that could be explained according to evolutionary theory. So a philosopher might well look askance at God's supposed omniscience and omnipotence, especially if our creation involved a mistake made on purpose, or a mistake that God was not omniscient enough detect.

As discussed in more detail in chapter 3, the general idea of intelligent design is that life is too complex, especially at its lowest level (cells), to have evolved on its own from nonliving materials. In this case, we don't really know what is meant by the term "intelligent" (as in "intelligent design") except as a relatively vague reference to powers that humans do not have, powers that allowed a supernatural being to develop particular molecules and arrange them in packages that exhibit and perform all the functions of life. Those life functions are: self-replication, response to environmental conditions, gathering of energy, processing of molecules (= metabolism = breaking and building molecules), exchange of materials with the environment, and, according to the scientists, evolution. Nobody ever argues with the first five of these properties; we see them manifested continuously among our fellow humans and, if we own pets, among those animals, too. If we garden, or manage a farm, then we see those same properties manifested in plants. The arguments over creation involve the last fundamental property of life, at least as recognized by the scientists—evolution.

Many religions, especially those surviving into the Third Millennium, don't deal very directly with creation issues, thus don't inspire people to write volumes on the issue of God's intelligence relative to bacteria and human beings. All religions have creation myths, of course, but rarely do such myths take on an aura of observed fact. Instead, creation stories retain their religious power by remaining myths, thus functioning as lessons about one's spiritual relationship to forces of the universe. Such lessons may not always be necessary for one's day to day existence, but humans are by their very nature curious and thoughtful, so our relationship to the forces of nature is clearly something we can and do wonder about. If we don't understand those forces, for example, volcanoes, hurricanes, and infectious disease, then we can easily attribute their power to a supernatural being. And for a culture that survives perfectly well without truly understanding how hurricanes develop, a creation myth can remain a mysterious but beautiful story instead of becoming a fact and it can help explain one's existence on Earth without members of that culture taking the myth literally.

Myths therefore can help us deal with natural disasters and collective misfortunes by blaming them on an angry God then changing our behavior in an effort to placate that God. No sooner had Hurricane Katrina devastated the Gulf Coast than Repent America, for example, a Christian fundamentalist group from Philadelphia, declared that Katrina was God's punishment for the sins of New Orleans citizens, especially tolerance of homosexuals. If the next Category 5 hurricane doesn't arrive for a fairly long time, then it's possible to attribute that fact to our successful placating behavior, and because we're pretty observant animals and can see our fellow humans behaving badly, the next hurricane can again be viewed as punishment, whether for the same sins repeated or different ones, or even sins that over the course of time have become sins because of technological innovations, child porn on the Internet being a good example of the latter. The real sins, of course, for which Katrina's devastation was payback, were those of the politicians who ignored, for decades, the readily available scientific knowledge about loss of delta lands' buffering capacity and the inadequacy of levees.

Hurricane Katrina may be fading from the American conscience, but the idea of blaming an omniscient supernatural power for natural phenomena has not. For example, certain religious leaders in the United States of America have followed this exact same train of thought relative to HIV and AIDS. See the July 18, 2007, issue of the *New York Times*, Health section for the quote: "According to a 1986 Los Angeles Times poll, one of four Americans agreed that 'AIDS is a punishment God has given homosexuals for the way they live.' And the media have frequently quoted declarations by religious leaders that AIDS is God's judgment on homosexuality or drug abuse." Understanding of HIV epidemiology and molecular biology, however, turns the problem into a relatively straightforward, but complex, one of viral life cycles and evolution, with quite predictable reductions in incidence and prevalence resulting from rather simple behaviors intended to prevent transmission (for example, use of condoms).

Are we as a population, or as a species, more "intelligent" now that we understand how HIV is transmitted and why infection with the virus can produce AIDS? There is no definitive answer to this question, but we can determine whether we are *acting* smart in relationship to our acquired knowledge. In other words, we can observe whether as a society we are indeed behaving in an intelligent way, performing all the acts that we know will reduce the incidence of HIV. The real question then becomes whether those acts somehow violate our species' sense of how we *should* be behaving.

For example, do needle exchange programs for intravenous drug users seem consistent with our ideal social standards? The answer to this question is "no," and certainly not if you believe that such programs legitimatize or promote illegal drug use. But are needle exchange programs the intelligent thing to do? Based on the experience of municipalities that have tried it, he answer to this question is an unequivocal "yes" (see sources in References).

In the last page or two we've discussed hurricanes, epidemiology of HIV, and AIDS. What do all these topics have in common? They are dangerous natural phenomena, and when we come in contact with them, and get hurt as a result, then our reaction to the hurt depends to some extent, perhaps a great extent, on two things: first, our knowledge about these phenomena, and second, what we do with that knowledge. In other words, our reaction depends quite a bit on how intelligently we behave.

Thus the main principle outlined in this chapter is illustrated by some rather familiar phenomena—hurricanes and HIV—although the illustrations could easily have been chosen from a long list of others—earthquakes, fire, influenza virus, cholera, being a few examples. That main principle is fairly simple: *intelligence is demonstrated by knowledge and actions based on knowledge*. But a second principle also quickly emerges from any serious consideration of intelligence, and that is: *intelligence is discernable only within the context in which it is displayed*. And there is still a third principle: *intelligent varies continuously over a fairly wide range, no matter what the context.*

A major issue relative to intelligence also involves the behavior of an individual vs. the behavior of groups. There is little doubt that compared to their fellow humans, some *individual* human beings are extremely intelligent in a variety of ways; anyone can find abundant evidence to support this assertion in any reasonably sized library. Individuals plan for the future, accumulate wealth, seek education and employment, provide loving care to offspring, and sometimes produce beautiful art, music, and literature that in turn inform other individuals of the human condition. Of course there are plenty of individuals who do exactly the opposite of all these behaviors, too, but from a statistical perspective, at least, individuals' behavior reveals not only a great deal of intelligence, it also shows that *Homo sapiens,* taken one individual at a time, is capable of truly noble acts derived from calm application of knowledge, experience, insight, objectivity, and analytical skills to problems that need to be solved.

It is also readily apparent, again from massive amounts of literature present in any reasonably sized public library, that when it comes to intelligent behavior, groups are far less capable than individuals of making decisions that are in their own long term best interests. Alternatively, for some reason, groups might be far less inclined [than individuals] to act intelligently. Many authors have tried to address the question, for example, of why nations behave the way they do. Barbara Tuchman's *The March of Folly* is one excellent and easily accessible book on this subject; Jared Diamond's *Collapse* is another. In fact, modern nations seem to act in ways that could be described almost in evolutionary terms: they act on their perceptions of immediate benefit; the same could be said of snakes and snails. In fact, nations tend to act just like individuals tend to act unless the individuals learn or are taught to act otherwise; that is, unless these individuals learn the difference between good and evil.

When nations (groups) do act in a rational, long-term, manner, it is usually because of some insightful, intelligent, and perhaps charismatic leader, or at least an individual with certain leadership skills—tone of voice, stature, posture, construction of his or her sentences, underlying rationale, etc. Franklin Roosevelt comes to mind as an illustration of a leader making a nation behave like a smart individual; Mahatma Gandhi is another. Again, your local library is an excellent source of data to support these assertions as well as provide other examples.

Were Gandhi and Roosevelt intelligent? We can't answer that question; all we can say is that given the environment in which they lived and the work they performed, the results benefited their fellow humans. Was Hitler intelligent? All the evidence suggests he was not exactly dumb, but instead was easily deluded into beliefs that had little or no basis in reality (see John Cornwell's book *Hitler's Scientists: Science, War, and the Devil's Pact*). Thus Hitler is an excellent example of a scientifically illiterate elected politician who turned out to be quite dangerous (see chapters 15 and 16); in retrospect, we would not call him intelligent.

How *should* intelligence be manifested? This question is a very interesting one for two reasons: first, humans tend to think of themselves as intelligent beings, and second, because we also consider certain non-human species to have intelligence. I contend for the purposes of discussion that non-scientists, or even scientists who are not biologists, typically consider non-human species to be intelligent when those species behave in human-like ways, especially the more noble of human-like ways.

For example, elephants remember their friends and relatives and dogs act happy to see us. Dogs give us unconditional love and ask only for a food, water, shelter, and some simple treats in return; few humans would consider such behavior "intelligent" if displayed by fellow humans. If dogs learn to fetch a stick, or sit up and shake "hands," then they are exhibiting some human-like behavior and we consider them smart, or at least trainable. Cats would never fetch a stick, but because we have such a wide variety of human models to choose from, we consider them arrogant, self-absorbed, and in possession of us rather than the other way around, so explain this rather unintelligent behavior on the basis of human analogies and even joke about it. The vast population of feral cats occupying our cities, however, demonstrates pretty clearly that as a species, within an environment offering food, water, and shelter, cats do well regardless of the fact that they are arrogant, self-absorbed, and won't fetch a stick.

Thus we see from this short discussion of intelligence that the term can be used in a variety of ways and can have different connotations depending on the various contexts within which it is used and the subjects being discussed. In this sense, *intelligence* is similar to many of our words describing human traits, including those traits that we also see manifested in non-human animals. When used in the phrase "intelligent design," the word refers to creative powers not only beyond those demonstrated by humans, but also not discernable in other creatures. We then conclude that such power must come from sources other than ones we can discover and observe, so the source therefore must be supernatural. Throughout history, supernatural forces have been considered gods, although in the modern American context, when the subject is "intelligent design" the supernatural force is God, particularly the Christian God described in the Bible.

Should elected officials and other busy culture warriors consider life on Earth to be the product of a God? That question can be answered only within the context of some situation in which such warriors find themselves. As we will see in later chapters, one's desires relative to a group tend to fashion our beliefs and behaviors in ways that we don't often recognize or admit. A person who wishes to be in a position of power over fellow humans, and to attain that position by convincing other people that he or she is worthy of such power, must tell the masses what they want to hear, especially if he or she wishes to retain that position for any length of time. As any well-educated individual with even a smattering of historical awareness knows, "the masses" are not very trustworthy when it comes to acting in their own long-term vested interests but are completely predictable when it comes to acting on their own emotions and beliefs, neither one of which has to match the reality of life on Earth.

Fortunately, for politicians, there is a whole lot of wriggle room relative to one's belief's regarding creation. This flexibility is illustrated beautifully by the political history of creationism, in which one finds almost as many interpretations of the Bible as there are fiery ministers to provide those interpretations (see chapter 4 for a summary of this history). Clearly the creationists have never agreed on the details of creation, or on how to reconcile their beliefs with the discoveries of science. Unfortunately, however, there is not nearly as much wriggle room in the public's tolerance of elected officials, especially presidents, who are well-educated, articulate, and rationale; this small bit of tolerance often fades into nothing if those officials also are relatively honest, hard-working, problem-solvers instead of fear-mongers who seem to be saving us from something dire (think "weapons of mass destruction" and "Iraq.")

Education and rationality, particularly if such education contains a little bit of biology, tends to tarnish the reputation of Genesis as science, although such education may well increase one's appreciation for Genesis as a myth of enormous power and beauty. Thus we have the dilemma of presidential candidates: how to appear ignorant enough to get elected if you're not. The larger dilemma, of course, is what to do about America's love affair with scientific illiteracy once you actually get inaugurated. Read on.

2. What is Design?

> *D'Arcy's treatment of form, then, is generally illuminating, particularly when he tells us how pervasive are certain elementary forms—the unduloid and catenoid, as well as their simpler relatives the sphere and cylinder . . .*
>
> —Ruth D'Arcy Thompson (*D'Arcy Wentworth Thompson*)

"What is design?" is a very difficult question to answer not only because the term probably means something different to every individual on Earth, but also because people use the word in such a variety of ways. For the purposes of this book, let's agree that "design" means an established, more or less stable and predictable, arrangement of parts usually intended to accomplish some function(s). This definition requires a designer, someone who purpose-fully—and some would say artfully—arranges parts and dictates their relationships, the latter leading to apparent stability. Stability then leads to recognition; we recognize a coffee mug because of certain elements to its design, but we also recognize any modifications of those elements—handle shape, proportions, color, size, company logo—that might occur when someone designs a different kind of mug.

If this definition and requirement for recognition seem pedantic, almost academic, then just go down to your local Walmart, Target, or other department store, and study some category of appliance like coffee makers. Your first reaction is likely to be: look at all these different *designs*. Such machines are intended to make coffee, of course, and they all do, and most of us understand why they do, but they are also intended to be visually appealing to buyers and users. So now we have added visual appeal to function, stability, and recognition as part of our definition of *design*. And if the designer is really intelligent, the coffee makers are easy to use, what we call ergonomically friendly. That is, you don't spill a lot of water and coffee grounds when filling it, the filters are easy to put in and take out, and the pot feels comfortable when you pour yourself or someone else a cup of java. Plus, it looks nice sitting there on the kitchen cabinet. So *design* implies a combination of utility, visual interest, and pleasure, or at least the intent to produce such a suite of characters.

Do living organisms have these basic design properties of function, stability, recognition, visual appeal, and ergonomic friendliness? The vast majority of familiar ones do, especially larger ones that comprise most people's perception of life on earth: companion animals, domestic stock, urban forest plants, etc. We assume they are ergonomically friendly because they are familiar, even though most of us don't earn our livings messing with them, but in truth, they are not. You have to be reasonably educated about horses, for example, to play around with them comfortably, but I'll admit that most dogs and some cats are pretty easy to handle. If they weren't, they would not be such popular companions.

So I contend that yes, indeed, most familiar organisms seem "designed" for some "purpose" at least in the minds of humans. The species that we recognize as wild also tend to have visual appeal when seen on television or in the zoo, but they're often quite uncooperative, if not outright dangerous, when interacting with people, and most of us would be hard pressed to come up with a "purpose" for a black bear or an armadillo other than "to glorify God," and even that "purpose" seems strained when we're considering armadillos.

Thus we can ask the question: Are living organisms really "designed"? The answer to this question is "yes." The American creation vs. evolution culture war is fought over the issue of who or what did the designing. Most biological scientists readily accept the evidence that living organisms are "designed" by evolutionary forces acting on genetic variability; the term "design"—in quotes—thus assumes a far broader definition than it might in more human-focused or religion-dominated circles.

Darwin himself viewed agricultural research as a lengthy and relatively successful exercise in the design of living organisms by humans using the same kind of evolutionary forces that operated on non-domesticated plants and animals. Creationists who "accept microevolution" understand this utilitarian approach to evolution but those same creationists draw the line somewhere between agricultural research and the production of so-called "higher taxa." In other words, *we* can produce broccoli, Brussels sprouts, and cauliflower from a wild plant named *Brassica oleacea* using Darwinian principles, but only God can make dogs, cats, horses, and whales from a common mammalian ancestor, or from scratch, as He chooses.

People who cannot accept the scientific evidence for evolution also tend to view living organisms as being far too complex to have been designed by natural selection. People who have studied science in general, and evolutionary biology in particular, especially with a relatively open mind, see no problem whatsoever with the generation of living diversity by natural selection and all its variations, as well as by a number of other mechanisms that introduce genetic, and subsequent structural, diversity into populations of living organisms. But because like real wars, culture wars are fought by people, this difference in viewpoint is not likely to disappear. Instead, the difference probably will become more and more visible as our nation struggles with its changing (evolving) place in the world community. The reason for this widening gap between the creationists and evolutionists on matters of design is the reliance on evidence and rules of interpretation. I elaborate on this reason in a later chapter, but for now, it's enough to say that the origin of design, and especially the processes that produce it, are at the heart of this debate.

Whenever we read the word "design," then it's important to also remember that human beings are, if nothing else, designers. We design clothing, buildings, furniture, appliances, automobiles, airplanes, boats, jewelry, books, magazines, advertisements, brochures, carpets, interior spaces, bicycles, SCUBA gear, hockey sticks, golf clubs, and a bewildering array of weapons. We also design cities, railroad yards, meat packing plants, irrigation canals, hydro-electric plants, parks and playgrounds, golf courses, skateboard parks, slalom courses, freeways, eating utensils, mechanical pencils, laptop computers, flash drives, cell phones, portable music containers, clothes driers, and coffee cups. This entire book, of course, could be filled with a list of stuff we design.

This habit of designing, including the manipulation of existing designs, is so deeply embedded in our genes that it is one of our most defining traits as a species. Any reasonable biologist from outer space, studying us carefully, would conclude after only a very short time that our crediting God with design of a bacterium is so expected as to be surprised if we did not do it. To design something is *natural* for humans; it's as natural as all other bodily functions, including urination, defecation, language, and sex. So the attribution of life's "design" to a Designer, in whose image we are supposed to be made, probably is a behavior rather deeply embedded in our genes.

The philosophical issues surrounding use of the word "design" are rather complex, mainly because of the above mentioned multiplicity of definitions and uses. With our propensity for simplification, we typically reduce the word's definitions to whatever we wish it to mean at the moment. If our definition fits somehow, at least in part, to whatever others have in mind, then we've made some friends. The problem with this phrase "intelligent design" is that the people who use it usually have no clue what they mean by the term. Which one of the 10,000 species of grass designs are they referring to? Which one or ones of the known 400,000 different beetle designs are they talking about? How about the half million species of one-celled parasites that live in those beetles' intestines? Where they each designed specifically by the Intelligent Designer for life in that particular niche? Or would our friends who believe in ID prefer, instead, to simply say something like "the Intelligent Designer produced an orchid, a thing of enormous beauty, and that is enough for me."

If that's all the further we go in this discussion, then we've stated our position on the matter of design and our like-minded friends firmly believe that they understand completely what we are saying. But there are at least 20,000 *known* designs of orchids, and probably a lot of ones that have not been discovered yet. So even by choosing something that illustrates our underlying world view—beauty, delicacy, and purpose summarized in a flower designed by Intelligent Designer—we simplify nature to such an extent that we are superimposing our own kind of organization upon it.

A philosopher might pursue this last line of thought by asking the question: if there were no people on Earth to see and react to it, would there still be this orchid, of a particular species that a human would view as beautiful, delicate, and placed here for the glorification of the Intelligent Designer? The answer to this last question, of course, is "yes," but there would be no debate over where the orchid came from. We have plenty of tangible evidence that this planet has been occupied by tens of thousands of species we might consider glorious, beautiful, and delicate, as well as a bunch of ugly and vicious ones, had we been here to see them, and might still consider them so were they not extinct.

In the minds of many people, the idea of design is more or less the opposite of the idea of chance. This chapter therefore is an appropriate place to deal with the definition of "chance," too. As indicated above, the word "design" implies purpose, but "chance" implies an accident or complete lack of purpose. When creationists claim that something so complex and beautifully designed, say, as the human immune system, could not have arisen by chance, then they are not really using the term "chance" correctly. So what does an evolutionist mean by "chance"?

This question is relatively easy to answer: By "chance" an evolutionist means a mutation, a heritable mistake in the replication of DNA, period. Evolutionists contend that what appears to us to be design can in fact arise, and probably fairly easily, through mistakes in the replication of DNA, and especially if environmental conditions sort through those mistakes, eliminating some and retaining others. Evolutionists also maintain that neither the mistakes nor the environmental conditions require an intelligent designer, although the sorting process sets up some boundary conditions under which subsequent mistakes are then sorted into additional, and different kinds, of successes and failures.

It is important to remember, however, that to the best of our knowledge, all species, indeed *all species*, vary both genetically and phenotypically, that is, in terms of the outward manifestations of their genetic makeup. Within any population of a single species, be it wolves, foxes, bison, houseflies, mosquitoes, or the hundreds of thousands of other known bacteria, protozoans, algae, fungi, plants and animals, there will be genetic variation and such genetic diversity will be manifested in the organisms' phenotypes. "Outward appearance" is a phrase that can also take on many meanings, from something as simple and obvious as eye or hair color, to characters more subtle but nevertheless real, such as the ability to build a certain kind of protein.

This genetic variability is produced by a number of different mechanisms, but the ultimate source of it is mutation, that is, a mistake in the making of DNA or damage to chromosomes. Either kind of mutation must occur during the making of gametes, or offspring, in the case of asexually reproducing species, or it will be lost when the organism dies. On the other hand, it is fairly obvious from even a cursory study of living organisms that they make lots of mistakes during gamete formation, and that many of these mistakes get preserved in offspring, although most of them we never see, disappearing because of their deleterious effects on the zygotes produced with mutant sperm and eggs. But some variants survive; look around any airport waiting area and you'll see plenty.

Over the long term, and in retrospect, the sorting of variants into survivors and failures can *appear* to be progresssive, that is, leading toward some "goal" such as the making of a modern human being from a prehistoric human-like primate. And, the sorting under changing conditions can ultimately produce a great deal of complexity. Thus when creationists use the term "by chance" to discredit evolutionists, they are generally referring to the complete assembly of a structure in a single event, but when evolutionists use that same phrase they mean small mistakes sorted into successes and failures, with the successes defining the options for future mistakes, that is, establishing boundary conditions.

Certainly the Empire State Building, for example, could not have been assembled all at once "by chance," but it could have been assembled by sorting through all the options for putting pieces of masonry and steel into various relationships, with choices being made by success and failure at every step of the way. Maybe the Empire State Building is too complex an illustration. Perhaps a better example might be the first dwelling constructed from rocks. With the initial stacking of rocks, the first low wall would either have fallen down or remained standing; the environment in which successes were sorted from failures would be a group of smart primates stacking rocks.

In the culture-wars discussions of evolution, eyes, for example, often are considered so wonderfully complex and perfectly designed that they could not have arisen just "by chance." Any biologist will readily admit that eyes don't arise anew, fully-formed, "by chance." Instead, they arise from previously existing eyes or light-sensitive structures although extensive study of invertebrate animals is usually required before a person can understand why eyes, in all their various forms, can probably be considered one of the easiest organs for animals to have acquired through evolution. In fact, organisms ranging from single-celled algae to hawks have such light-sensitive structures, and these organelles and organs are made from diverse cell components and tissues, perhaps the strongest evidence that eyes are of evolutionary origin.

An instructive analogy to eyes in the animal kingdom might be some common human apparel such as hats. If hats were designed by an Intelligent Designer, you would expect them to have originated once, and you would also expect us to be able to trace the diversification of hats back through human history the same way we can trace micro-evolutionary changes in corn or wheat. Hats, however, resemble a number of structures in the natural world—for example, eyes—because they seem to appear, in a particular form, and made from particular materials, depending on their "need."

For example, a football helmet has quite a different origin and use from a dunce cap, (although certain fans might equate the two, depending on who is wearing the helmet), and thousands of similar comparisons could easily be made between various types of headgear, including military helmets, which originated centuries before American football was ever invented. Such independent origins of similar structures ("designs") that fulfill similar or identical roles, is one of the strongest lines of evidence supporting evolution as the source of biological diversity. One familiar example compares extinct ichthyosaurs and modern porpoises, each derived independently, occupying the marine environment, and evolving similar designs, both from terrestrial ancestors although tens of millions of years apart.

This chapter also is an appropriate place to deal with the idea of randomness. "Random" means without a predictable time or place of occurrence, although prediction in this case refers to an individual occurrence, not a pattern of occurrences. Thus we can predict that a mutation (an alteration of genetic information) will occur some time, but we cannot predict exactly when that mutation will occur or in which gene. With enough study, however, we can estimate the probabilities that certain events, including particular mutations, will happen. Thus we know mutation rates for many genes, failure rates for machinery, automobile accident rates, etc. From such rates, we can construct frequency distributions (a pattern of occurrence) that predict the number of such events per unit time.

With further research, sometimes we can also determine underlying mechanisms for randomly occurring events, as in the case of radioactive isotope decay or mutation. Your automobile insurance rates are determined, to a large extent, for example, by the pattern of occurrence of accidents involving drivers of your age, education level, gender, and past driving record. Nobody can predict exactly when a given accident will occur; otherwise it would not be considered an accident, but actuaries can calculate the probability of such an accident and adjust your premiums accordingly.

Mutation may occur at random, as in the appearance of a gene for fruit fly vestigial wings, but subsequent mate selection by flies, based on mating "songs" produced by vibrations of vestigial wings vs. normal wings, is certainly not random. Thus the most powerful constraint of all on evolutionary change is existing genotype and the resulting phenotype, that is, the boundary conditions for future change. From viruses and bacteria, to protists such as *Paramecium* species, to song birds, snakes, and people, change occurs only from an existing genotype (genetic makeup) and phenotype ("appearance," or an outward manifestation of genetic makeup), and occurs only within the limits allowed by that genotype and phenotype.

Stated simply, according to evolutionary principles, cats don't randomly evolve into dogs by chance, and birds don't evolve into people, either. Instead, cats evolve into different kinds of cats and birds evolve into other kinds of birds. But as long as we're talking about birds, remember that biologists now consider birds to be small dinosaurs (and for good reason). And there is plenty of evidence, both fossil and molecular, for a shared ancestry between cats and dogs.

Thus in order for us to understand evolutionary biology, particularly that part of it concerned with design, it is important to remember exactly what it is that occurs randomly or by chance, and then to put such random occurrence into the context of boundary conditions and constraints. It also is important to be aware of the individual variability we see all around us in all kinds of organisms, including ourselves. Is this natural process of sorting through variants powerful enough, over extended time, to change organisms from one species into another, or to produce diversity from relative uniformity, that is, to generate families of organisms such as mice and mosquitoes, from ancestral mammals and insects, respectively, that looked nothing at all like their descendents? Evolutionary biologists would certainly answer this question in the affirmative. There is simply too much evidence that the process works, and most of that evidence comes, as Darwin recognized, from agricultural research and breeding of companion animals.

The issue that stirs up Biblical literalists, however, is whether the process of natural selection is powerful enough to generate so-called "higher taxa," for example, to produce both horses and pigs (odd- and even-toed ungulates) from ancestral stock, to produce both whales and hippos from a common ancestral stock, or, of course, to produce both humans and chimpanzees, also from a common, non-human, ancestor. Is this natural process of sorting through variants, resulting in permanent change, operating today on human beings? Evidence supporting a "yes" answer is relatively strong. In this modern case, as in all previous and known cases involving non-human species, the sorting is being done by means of reproduction. Although clearly many humans die before achieving reproductive age, all current population estimates and birth rate data show clearly that our species is becoming smaller and browner.

Translated, "smaller and browner" means "Indian, African, and Asian" The names that tended to dominate American news in the opening years of the Third Millennium—with Vladimir Putin and George W. Bush as prime examples—are not small and brown, and this fact alone diverts our attention away from what is happening to our species on a global scale. We *Homo sapiens* are being designed, and fairly rapidly, as a result of differential reproductive rates in various areas of the globe. Is such designing by forces over which we have theoretical, but no practical, control a bad thing? No; it is a natural thing. Will we still be recognizable as human beings two millennia hence? Yes; of course we will, assuming we have not become extinct, as have so many other species, by our own hands.

3. What is Intelligent Design?

> *Three times at least, but probably much more often, eyes with lenses have evolved independently in animals as widely different as mollusks, spiders and vertebrates. . . there is no limit to what an artist can do with Gozzi's meager list of thirty-six themes.*
>
> —Arthur Koestler (*The Ghost in the Machine*)

The term "Intelligent Design" as used in the culture wars refers to the idea that life is too complex to have been produced except by a superior intelligence, that is, God. In some versions of the idea, not only life, but the entire universe is considered too complex to have simply appeared in a Big Bang. So "Intelligent Design," hereinafter called ID, is a nascent, quasi-scientific, theory—"nascent" because it has not withstood the repeated testing that full-blown scientific theories receive (except in the realm of politics), and "quasi" because this theory gives rise to no testable predictions (except, again, in the realm of politics). There is no such thing as "creation science" unless one considers all the bio-chemical research on the origin of macromolecules to be such. Religion gives rise to no testable predictions about the natural world beyond those involving human desires and beliefs. To its credit, religion has plenty to say, and plenty of predictions to test, about desires and beliefs, but none to test about the origin of biological diversity on Earth or on any other planet in the universe.

ID can be considered not only an idea and a quasi-scientific theory, but also as a world view, a personal device for establishing the nature of reality, an historical phenomenon, a means of competing for political power, a relatively undefined realm of intellectual endeavor, and a vocabulary word (phrase) that condenses a large body of religious thought into a single negative one (ID = not evolution). In culture wars, condensation equates to simplification, and in most cases, the resulting simplicity is a reflection of willful ignorance. Most natural phenomena are quite complex, as are most hot-button political issues. Immigration and abortion are perfect examples of social phenomena that politicians can, and readily do, simplify for self-serving purposes ("Deport those who are in this country illegally!" "Repeal *Roe vs. Wade!*") Viewed in historical and biological contexts, however, both these hot buttons are exceedingly complicated phenomena that have multiple underlying causes, permeate our society, and have major social and economic impacts no matter how they are manifested.

World views are often helpful to individuals, which is why they are also useful to those seeking power. If you have a picture in your mind of how the universe operates, then that picture relieves you of a whole lot of reading and research; in other words, you don't have to worry or wonder about where the stars came from. But that picture can also be quite polarizing, as the culture wars so clearly demonstrate. We tend to gravitate towards those whose world views match ours, and we tend to give them credit for being more astute than they are, especially when these folks are in positions of power (elected officials, priests, etc.). World views derived from religious beliefs are especially vulnerable to co-option by individuals seeking to acquire or retain power. A check of your local daily newspaper will provide abundant evidence, on a global basis, to support this last claim.

Like all ideas and beliefs, ID has a history, too. That history actually started with the Classical Era Greeks, including Plato and Aristotle, and continued in various forms through Victorian England to 21st Century United States, sustained by a variety of forces, including wonder and belief. The history of creationism is summarized briefly on Wikipedia, which probably the most neutral site among the nearly four million hits one gets with a Google® search using "intelligent design" as the key word. An organization known as the Discovery Institute houses the most vocal and dedicated of ID proponents. The institute describes itself as "a nonpartisan public policy think tank conducting research on technology, science and culture, economics and foreign affairs," but it is anything but non-partisan on matters of science, especially evolutionary biology. A quick check of their slick web site reveals why their non-partisan claim is disingenuous, with news postings such as "The Dark Side of Darwinism" and "Hawking Irrational Arguments."

ID history, along with that of creationism in general, is certainly worth studying, if for no other reason than it reveals our species' willingness to fight over ideas, to believe charismatic authority figures, and to pass judgment on others based on their religious beliefs. Ronald Numbers, eminent scholar from the University of Wisconsin, has summarized this history eloquently in his book *The Creationists: From Scientific Creationism to Intelligent Design* (mentioned elsewhere), which I recommend strongly as evening reading.

Historians in general, and especially established scholars like Dr. Numbers, are often better equipped to analyze various intellectual movements than are the participants in those movements because they spend their careers delving into primary sources (correspondence, court records, etc.) and answering to anonymous peer reviewers. Furthermore, historyians usually are more interested in the way various events shape the course of human affairs than in doing the shaping themselves. Thus I tend to trust the historians far more than folks with agendas, including, for example, those "scientists" representing the ID movement via the Discovery Institute.

ID is a relatively undefined realm of intellectual endeavor because there are so many possible interpretations of the actual events that supposedly occurred during the design period. If we assume there is a Supreme Being who is eternal, omnipotent, and omniscient, that is, lives forever, can do anything, and knows everything, then our biggest challenge as human thinkers is to avoid bestowing this Supreme Being with obviously human characteristics. I tried to accomplish this avoidance task in a small book entitled *Conversations Between God and Satan, Held during October, 2004, at The Crescent Moon Coffee House in Lincoln, Nebraska, USA, Earth, Milky Way* (www.createspace.com/3431482). It's actually pretty amazing what surfaces in your mind when you assume eternal life, omnipotence, and omniscience. Much of that mental bubbling up concerns life in other galaxies, and even consideration of this possibility puts you at odds with large segments of the conservative Christian community. So the fundamental question, stemming directly from the idea of ID, is: what did God really make, and where did He/She/It make it?

Answers to these questions are beyond the capacity of anyone on Earth to answer satisfactorily. Even the concept of irreducible complexity, a favorite with scientists who argue in favor of God's existence, when considered objectively, turns out to be just one interpretation of nature, and a not very accurate one. Michael Behe, author of *Darwin's Black Box: The Biochemical Challenge to Evolution*, is the major proponent of this concept, describing irreducibly complex systems as being "composed of several well-matched, interacting parts that contribute to the basic function, wherein the removal of any one of the parts causes the system to effectively cease functioning."

Behe is describing cells, but he could just as easily be describing a basketball team. The scientific community has been highly critical of Behe, as one might suspect. The negative criticisms are relatively well-founded, but like most if not all modern research in biochemistry, biophysics, and molecular and evolutionary biology, the technology involved, the resulting observations, and the valid interpretations are far beyond the ability of an average, even well-educated, citizen to understand, much less argue intelligently about.

A typical reaction to highly technological information, at least among Americans, is "if I don't understand it, then it must be wrong." For "wrong" you can substitute "unproven," "unsubstantiated," "not important," "bad," "bad for me," or "just a theory." Americans are not very patient with things they don't understand, especially if that lack of understanding involves beliefs, particularly religious beliefs. So the arguments about ID and irreducible complexity remain best conducted among scientists who understand the physics, chemistry, and math involved, or among lawyers dealing with cases such as *Tammy Kitzmiller et al. vs. Dover Area School District, et al.* Attorneys, probably including many who squeaked through their college liberal education science classes with great impatience, nowadays often find themselves on one side or another of technologically complex cases. The one from Dover, Pennsylvania, certainly falls into that category.

A full history of this case is readily available on the Internet to anyone in the world with electronic access. The judges' conclusion was that ID is clearly creationism, that is, religion, cloaked in delicate and scholarly language. Remember that the scientific community arguing against ID is the same scientific community that brings us all of the research, discoveries, and technological developments that pervade the modern world, including realms of medicine, business, communication, agriculture, and war.

The phrase "science is a two-edged sword" is old, hackneyed, and trite, but also true. When humans study the universe, and the processes that occur throughout that universe from distant galaxies to the inner molecular workings of cells, then they make all kinds of discoveries, some of which tell us things about ourselves that we don't all want to hear. Good examples of such discoveries are ones such as (1) we obviously evolved from non-human primates, and (2) the origin of life from non-living materials is not all that implausible, given what we know about the abiotic formation of complex molecules and membranes.

As a political weapon, however, ID is a well tested and reasonably successful one. That is, ID works to polarize people, giving them a sense that they are being attacked, or persecuted, for their beliefs, and that such abuse comes from people who do not believe in God. God, whatever He/She/It is, obviously stands on the side of believers and God is always right. You can't argue with eternal omnipotence and omniscience, so a large number of us seem unable, or at least unwilling, to argue with the idea of a Supreme Being.

Political weapons always work best when they inspire feelings of persecution, betrayal, and abuse, and thus function to bind the persecuted into a tight-knit group. Religion has always, throughout all recorded history, functioned very well to polarize people, make them feel threatened and defensive, then act on their feelings. ID is highly effective in this regard, especially because it clearly delineates between "us" (believers in a higher, supernatural, power) and "them," the evolutionists, atheists, secular humanists, Marxists, Communists, pro-choicers, homosexuals, and probably illegal immigrants, too.

Like other political weapons (hatred of taxes, imaginary Communists, socialized medicine, etc.), it's entirely possible to develop a theoretical framework for their use. In the case of ID, the theory takes into account the fact that as a species, humans are quite gullible and superstitious, and generally believe in super- or extra-natural phenomena such as gods, angels, ghosts, or spirits. This trait may well have a biological origin, arising as some kind of an emergent property in the complexity of our central nervous system. Richard Dawkins, in his book *The God Delusion* suggests that religion may have an evolutionary origin, functioning to help toddlers respect authority and binding primitive people together into bands of cooperating individuals. As you might suspect, Dawkins is quite high on the ID community's anathema list and his book produced a firestorm of negative reaction from the religious establishment (again, see Wikipedia for quick access to this body of information).

Regardless of their ultimate origins, the human brain and its ethereal companion the mind, are rather extraordinarily complex information processing systems. Highly complex systems, especially those involving information flow, typically give rise to all kinds of emergent properties—rumors, conspiracy theories, reconstructed history, etc. We humans believe strongly in the supernatural (for example Heaven and angels). A Google® search using "contact with deceased loved ones" as the key word(s) turns up nearly a million hits. A political theory of ID might well begin with this very human trait that allows us to enter imaginary worlds through our minds, then take into account our fear of the non-understood, and finish with goals to be accomplished. Most well-educated scientists believe that the long-term goal of ID proponents is institutionalized willful ignorance.

The fact that ID gives rise to no testable predictions, however, means that no matter how effective it is as a political weapon, it will never achieve scientific respectability, no matter what fraction of the general public believes it fervently, no matter how much political play it gets, and no matter how much it gets incorporated into public school curricula in backward communities determined not to sustain or develop their children's scientific literacy in a technologically competitive global economy.

Thus we have a concept—that Earth and its inhabitants are too complex, and finely adapted to a highly specific set of conditions, to have arisen except by design of some supernatural power—a concept that functions to help erode the very human resources that allows that human population to survive with any sense of dignity. When a great and powerful nation, whose economy is sustained by highly technical and sophisticated endeavors, chooses to be ignorant about the natural world, then that nation is in long-term trouble. ID, and some would claim religion in general, is a significant factor in this trouble.

ID sometimes gets support from people with strong mathematical skills. Such scholars typically are keen to point out that the probability of these complex structures arising "by chance"—eyes being oft-cited examples—is virtually zero. Thus in the minds of these mathematicians, this calculation precludes evolution, which is interpreted as a process in which structures arise "by chance," and so infers a designer. The problem with this kind of an argument is that mathematicians are not biologists, and especially not invertebrate zoologists, and rarely do they understand the first thing about evolutionary theory or about organisms. Neither do physicists or medical doctors who also claim that eyes are too complex to have arisen except by the Hand of God.

Such claims leave invertebrate zoologists scratching their heads. If there was ever a structure that has arisen repeatedly in diverse groups of animals over the past 500 million years, and thus by inference rather easily (in evolutionary terms), it is eyes. Furthermore, the embryological development of eyes is one of the more transparent clues to evolutionary history provided by ontogeny (long a commonly accepted indicator of fundamental relationship, thus of evolutionary history). So eyes are not a very logical or satisfactory illustration of structures too complex to have arisen by chance alone.

Having said all of the above, however, it is important to remember that it is extremely unlikely for eyes suddenly to appear full blown and functional in some group of animals that has never had them. Such an event would be the rough equivalent of the Empire State Building appearing just as suddenly, assembled as if by random encounter between bricks, mortar, glass, and electric cables lying around on Manhattan streets, with no apparent forces to produce the structure. Admittedly, if the Empire State Building were the only building on Earth, and we were still walking and riding horses, killing wild animals and collecting berries for food, instead of zipping around in cars and stopping off at the supermarket, then it clearly would seem like the work of a supernatural power. But it's equally important to remember that vertebrate animals have had eyes for at least 400 million years, and invertebrates for even longer.

It is only with an understanding of the history of architecture that the Empire State Building makes sense. It is no act of God; it is a magnificent but not particularly creative piece of Depression Era Art Deco urban construction. And, if course, the Empire State Building had a designer; his name was Gregory Johnson. Along with his architectural firm—Shreve, Lamb, and Harmon—Johnson produced the designs in about two weeks. Sources from architectural history suggest that Johnson may have borrowed heavily from existing designs, but nevertheless he is given primary credit for the building.

If we were still back in that hunter-gatherer stage of social-cultural development and Gregory Johnson showed up at one of our clan meetings with a set of blueprints for the Empire State Building, trying to explain exactly how this project would be assembled, and from what materials, we would think he was absolutely crazy. Or maybe we would consider him God. Or perhaps one of several Gods. Or we might just kill him for being so different from us as to be considered a threat to the public order. Because of the constraints of our history and culture, we'd never realize that he was just another human being with a lot of talent and education.

But life is not architecture, and we have never created life *de novo* the same way we have built tall buildings. Humans have, however, done a lot of experiments in which molecules get assembled without the hand of God, or, if God is at work, then by His work done according to well-established chemical principles. Granted, there is a great distance between an amino acid that is formed in a closed glass container subjected to electrical sparks (do a Google® search on "Stanley Miller experiment") and a living, moving, eating and excreting ameba. But there is very little doubt that the fundamental building blocks of living organisms, namely the complex carbon-containing molecules, can easily be formed from smaller molecules we know are common throughout the universe. Thus the hand of God must be relegated to subsequent assembly of these molecules into still more complex ones whose *arrangements* and *sustained interactions* breathe true life into a system.

So "intelligent design" now becomes God's use of preformed molecules—methane, carbon dioxide, and water, to mention a few really common ones—to make living organisms. To my knowledge, nobody in the intelligent design community has argued that planets, stars, and interstellar gasses are too complex not to have been made by God, although there is plenty of argument that the entire universe is "just right" for its physical laws to produce the diversity we see: a multitude of stars and star types, planets of various sizes and compositions, dust and gas clouds, asteroids, etc. Politically, however, ID is focused on familiar animals, especially members of the species *Homo sapiens*. Most of us don't care a lick about possible amebas in mucky ponds on an undiscovered planet orbiting some star out in the Large Magellanic Cloud. We care a lot about our relatives and our children; none of them look like apes to us regardless of how apish they may be acting at the moment.

I contend that a reasoned, calm, and relatively deep consideration of exactly what a God must have designed brings one to the conclusion that that the very idea of intelligent design is fraught with problems. I am certainly not the first, nor will I be the last, to point out such problems, nor will my potential critics—if they even notice what I've said on the subject—be the firsts and lasts to counter any of statements in this book. Nevertheless, my assertions and the critics' counter assertions are strictly, and typically, *human* products, participating in *human* endeavors, namely, intellectual politics. We humans cannot discover where the universe came from, but we can make up stories, some, like the scientists', based on observations about nature. The claim that just because we don't know where the universe came from means that God must have made it, is exactly the same kind of claim that the ancients must have made when they named the constellations.

In essence, therefore, ID is a nice idea that helps organize a massive realm of knowledge into something a large number of people believe they can understand. Furthermore, ID is consistent, sort of, with a wide range of myths, a situation that seems to validate it as an explanation for why we exist. But ID also is a political weapon that can be used to destroy rationality, especially rationality developing in the minds of school children, and for that reason it is to be recognized as religion, period, not science. A review of history quickly reveals that no matter how useful religion is to individuals trying to make their peace with whatever circumstances the laws of chance have delivered to their doorstep, it is nevertheless a major public health hazard when it infects governments. ID may be no more dangerous than a case of metaphorical flu, but remember that common flu viruses are dangerous as hell to those whose immune systems are somehow flawed or compromised.

4. What is Complexity?

> *Psychologically, one of the great powers of programming is the ability to define new compound operations in terms of old ones, and to do this over and over again, thus building up a vast repertoire of ever more complex operations. It is quite reminiscent of evolution . .*
> —Douglas Hofstadter (*Metamagical Themas*)

When scientists use the word "complexity," they usually are referring to a situation in which there are many participants and many diverse kinds of interactions between those participants. "Many" is a relative term, but any time an activity involves more than three participants and three to five types of interactions, then that activity starts to appear relatively complex to the average person. A family with two working parents and three teenage children comes immediately to mind, although if two of the children are pre-teens, with soccer and music lessons on top of school, then a typical week's schedule can seem like the very model of complexity.

A professional basketball game is several orders of magnitude more complex than a week in the life of a typical suburban nuclear family, and if you add the off-court machinations of both teams, then the system (= teams + league + fans + media + money + contracts + history + potential draftees + criminal behavior) becomes very complex indeed. But compared to what goes on inside a micro-organism such an ameba, or a square yard patch of tropical forest, both the nuclear family and the sometimes seemingly nuclear (but in a different sense!) NBA are exceedingly simple systems built of a relatively few parts and maintained by participants' interactions.

Why is it important to address early on this particular question—What is complexity? I offer four reasons, the first being that in one of the United States' most persistent, pervasive, and unflattering cultural conflicts, namely, the so-called battle between "creation" and "evolution," the most recent, and probably most successful, of various anti-evolution movements is called "intelligent design." The term "intelligent design" actually refers to an idea that the universe is much too complex to have arisen without an "Intelligent Designer," that is, a God. This complexity is manifested especially in living organisms on Earth, with special reference to cells and their internal reactions and thus by inference, large familiar animals such as dogs, cats, and humans, too. But as in the case of "evolution," "intelligent design" as used by the media has come to include the ideas, all the writings, all the people who believe in it, and the political activities of those people. In other words, "ID" is a very simple abbreviation for a highly complex phenomenon that has evolved into many different versions.

The second reason to address complexity is that nowadays, at least in what we call the "developed world," that is the United States, Europe, Japan, and much of Asia, people tend to be inundated with information—literally thousands of messages daily from advertising, radio, television, newspapers, and other people. So anyone who tries to get through this communications overload has to make a message quick, simple, and aimed toward the emotions, a good example being a cell phone text message that reads "i love u lets do lunch." Thus in the face of high complexity (the information overload), or potential complexity (i love u), we retreat into simplicity (lets do lunch). The absence of correct grammar and punctuation in this example is typical of the medium.

Such simplicity does not make complexity go away; instead, we assume we've dealt with that complexity whether we've done so successfully or not. The letters on your smart phone's screen, for example, eliminate the sender's tone of voice, whomever he or she might be holding onto at the moment, his or her location (depending on your hardware and software), and his or her state of mind. You the recipient of this simple message also have the power to interpret it however you want to. Your options range from erotic excitement to an instantaneous deletion and anger. But when the living, breathing, warm, in-person human dimension is added to this simple message, it can become quite complicated, especially if it's filled with duplicity, nuance, and relativity. If the sender shows up and delivers this message in person, then you suddenly discover that the cell phone has stripped both you and the sender of your humanity. In other words, the live animals in their natural environments establish the context for those six words, and the context complicates matters rather substantially.

The third reason for addressing complexity is that in the United States, politicians are notorious for simplifying extremely complex social issues, abortion and gay rights being superb examples. You're either for them or against them, period, which is many places translates into for or against particular politicians, which in turn probably means lose or win the next election, respectively. Extreme simplification of emotional, highly charged and complex social issues by politicians does not make these issues less complex. Instead, such simplification delays resolution of social conflict by maintaining it, and in some cases fostering it.

For example, there probably are as many different reasons for someone getting an abortion as there are women who get them. Likewise, there probably are as many different kinds of homosexual people as there are people in general because homosexuality is rather like the hiring practices of most American government agencies in the sense that it tends to be pretty evenly distributed regardless of race, religion, gender, age, or country of origin. The last two sentences should be a warning for anyone listening to anyone else trying to simplify these two hot-button issues for political gain.

But the fourth and last reason for dealing with this question—What is complexity?—is most important: the universe is extremely complex at both very large and very small scales. So unless you're tolerant of complexity, or at least willing to try understanding a little bit of it, you're going to remain a hopelessly ignorant soldier in the culture wars. In general, ignorance is dangerous, but especially so in warfare. Does a culture war qualify for an activity in which ignorance is dangerous, even though there is not much shooting and bombing involved? Personally I believe the answer to this question is "yes," mainly because cultural conflict can quickly devolve into political decisions that influence many lives, although it is also well to remember that in many parts of the world there is indeed shooting and bombing that can be interpreted as a result of cultural conflict. Ignorance also makes people do things that in retrospect seem pretty dumb, although at the time they do these things in their state of blissful ignorance nobody realized the behaviors were so stupid. Hindsight, goes the tired old saying, is always 20-20.

Thus knowledge and understanding have a way of putting our past actions into a new perspective. In addition, the rest of this little book is not likely to make much sense unless we're all on the same page relative to complexity. So my advice is to sit back, relax, and let your mind wander through the following arguments, assertions, and ideas. Although there is quite a bit at stake in the culture wars, you're not in any physical danger, at least not yet, in the United States, by being a participant, regardless of the danger your nation may be facing because its elected leaders simplify political issues that are extremely complex on a global scale. But be forewarned: my definition of complexity is relatively simple because I'm more concerned with your general understanding of its fundamental nature than with the extent of your knowledge of specific cases. You can always look up the specific cases and study them in depth.

Remember that *complexity* is usually defined as a state or condition in which multiple parts interact in a variety of ways. Any team sport, especially as played at the professsional level, is a great vehicle for learning about complexity in general. A massive body of literature exists, much of it readily accessible in any public library, to support this assertion, but I strongly recommend the *Sports Illustrated* 2005 NFL Preview issue, because of the articles on development of a professional football team's playbook. Such tomes make your average advanced biochemistry text look rather simple, but when we see all the theory, good intentions, and training presented as a product on our TV screens, our reaction is quite simple: the players did or did not make [enough] yards. The same could be said of an individual sport like golf; if it were easy (simple) to hit a golf ball far and in exactly the direction we want it to go, from any kind of grass or hillside, then we'd all be scratch players. Instead, the opposite is very true. Hitting a ball with a stick is an exceedingly complex activity much like war in that it has many potential outcomes, most of them undesirable.

Human beings have a rather amazing capacity for simplification, and other species may, too, although we really don't have much information on their mental processes. To us, a cheetah's behavior might be described simply as "run fast catch antelope." Deep inside that cheetah's body, however, decisions are being made (mostly without the cheetah's knowledge or awareness) about everything from the contraction of individual muscle cells to the chemical reactions by which hydrogen and oxygen are bonded to make water.

To our credit, we could probably not function in the real world unless we were able to simplify complex activities such as killing a wooly mammoth into an easily understood phrase something like "get food find cave." Somebody at the mammoth-killing stage of our cultural development, however, saw beyond this simple practical requirement, picked up red dirt, ground and mixed it up into paint, crawled hundreds of yards back into the cave, chose a particular wall to use as an easel, and made a picture of what it actually meant to kill a mammoth. Fifteen thousand years later we are still trying to unravel the complex social and intellectual processes that led to these pictures. Complexity does not yield its secrets easily.

How is complexity created in nature? This question is, of course, at the heart of any "evolution vs. creation controversy" and to answer it we might once again consider the socalled nuclear family, also known as the archetypical 50s era TV family. The easiest way to create complexity is to add components to a system (see the Hofstadter epigraph at the beginning of this chapter), especially components that must interact with existing parts of the system, while retaining evidence for what the past was like. A 50s nuclear family is our perfect model. A new baby is an addition to the system. Nobody in this family forgets the history, and all are influenced by that history. Parents remember what it was like before they had children, and pictures probably help sustain this memory. Grandparents die and parents treasure the pictures of those grandparents holding grand children. Parents eventually get old and sometimes look back through scrapbooks at holiday cards, birthday cards, athletic prizes, high school prom photos, and notes from teachers.

When parents die, children come in to clean out the old house. They never question where all this complexity came from because they understand the processes of accretion, change from existing conditions, loss, and substitution, and they not only recognize, but easily interpret, evidence for the past. The children find their own baby pictures, taken way back when, and marvel at how this tiny creature eventually became an expert evolutionary biologist, albeit one whose total body of expertise is applied only to a single human family. No culture warrior walks into the sad household, looks around at all the junk and knick-knacks, and proclaims this mess was too complex to have arisen by chance. The grieving children would tell this jerk that whatever "chance" was at work to build this home, its contents, and its memories, that "chance" was highly constrained by boundary conditions, choice, and context over the decades it took to assemble the entity now called "our dead parents' house."

So this principle—addition of new parts with retention of old parts, or at least of evidence for what old parts were like—is at the center of our discussion. In the case of our evolution vs. creation culture war, however, the parts are those of living organisms, and the evidence takes a variety of forms including fossils, genes, etc. At the outset, we must all admit that no humans were present when life first appeared on Earth, so we really do not know, in the same sense that we know our morning paper was delivered by the kid up the street, just how those first organisms were made. However, we also must admit that in the case of murders, the perpetrator is usually long gone from the scene and the victim cannot tell us how this crime occurred. Instead, we, meaning the people represented by law enforcement agencies, must reconstruct the past from available evidence. Eventually, if someone is arrested, then this evidence must be presented to a jury, usually made up of taxpayers and registered voters whose education about law enforcement comes from television and movies. If someone is to be convicted, then these people must be convinced that a prosecutor has reconstructed the past accurately. Because crime is the foundation of so much of our entertainment, we are quite familiar with the act of reconstructing the past.

Evolutionary biologists also spend their days reconstructing the past, in a manner somewhat analogous to that of a detective. Indeed, such scientists are detectives in a very real sense. They must find evidence, reconstruct situations, seek motives (= causality), and infer process (how something came about). Is a single bacterium too complex to have arisen from non-living materials? This question is the central one of the evolution vs. creation battle. Once bacteria are present on Earth, then the evolution of all so-called "higher" forms of life, including us, becomes much more plausible than it otherwise would be, at least from a scientific viewpoint. So the real issue in the ID vs. evolution culture war is whether anything at all like a bacterium could have self-assembled out of simple molecules then acquired the properties we associate with life, namely the ability to capture and transform energy, then use that energy to replicate.

The scientific community claims this self-assembly not only is plausible, it also is the best explanation for our existence; the ID community claims the opposite, and indeed asserts that self-assembly of such a complex entity as a bacterium is impossible. The scientists seem to win this argument in court because the methodology of science allows one to gather legitimate evidence to address the problem of how life arose. The ID community loses in court because their solution to this problem is to claim "God did it," and there is no scientific methodology to use to test the assertion of supernatural intervention. As in a criminal case, methodology is at the center of any official decision, especially by a court of law.

We need to admit at the outset that in order to argue creation vs. evolution at the professional level, people should be extremely well educated in biochemistry, mathematics, physics, philosophy, rhetoric, and history, especially historical research. Few people have this kind of broad, almost Renaissance-type, education. Indeed, biochemistry, mathematics, and physics are, today, such complex and difficult subjects that professionals routinely hide in their subdisciplines rather than confront the complexity of their own broad areas of expertise. Nevertheless, my goal in the next couple of pages is to provide a framework for understanding arguments regarding complexity of life at its presumed origin.

If you need additional information about the chemistry involved, that background is fairly easy to obtain, at least in summary form, through Wikipedia. Most of my colleagues cringe at mention of Wikipedia, but it's cheaper than a PhD and it's at least a place to start. But at the outset you do need to know and be able to use some vocabulary words, specifically: *carbon, oxygen, hydrogen, nitrogen, sulfur, iron, mole-*

cule, methane, carbon dioxide, and *ammonia.* These words refer to the most basic building blocks of life. They also refer to substances that occur both throughout and outside of the Solar System and throughout the universe beyond. Nobody in the ID community disputes either this observed fact or the methodology behind the observations. The question of origin addresses the assembly of these substances into cytoplasm that is capable of garnering and transforming energy and replicating itself.

The scientific community claims such assembly is possible; the ID community claims it is not. In their arguments, scientists invoke a variety of environmental conditions and known processes; the ID community says these conditions and processes are either ineffective or irrelevant. There is no way to resolve this debate. There is, however, plenty of room for research on the processes and events that occur under certain conditions, including those that seem to mimic primordial Earth, and if it accomplishes nothing else, the scientific community does research, period, and on just about anything that stimulates human curiosity, including the origin of life.

There is a scientific journal devoted to publishing research intended satisfy scientists' curiosity about our origins, and its name, not surprisingly, is *Origins of Life and Evolution of the Biosphere*, although there is also a great deal of very sophisticated origin research published in some of the other widely respected journals. A crucial element of this discussion of complexity and life's origin is the basis for respect relative to scientific publications.

In general, scientific journals earn respect through anonymous peer review of articles submitted. The more harshly research is criticized, the more detailed its review prior to publication, and the more people trying to publish in a particular journal, then more widely respected that journal tends to be. What constitutes anonymous peer review? In the vast majority of cases, experienced scientists examine methods, including experimental design, very carefully, along with the materials used, the statistical analysis of results, and the author or authors' interpretations. Just like a skilled carpenter can tell whether a cabinet is built correctly, a skilled and experienced scientist can tell whether another's work has been done correctly. Finally, the essence of science is repeatability. If experiments cannot be repeated by others, with the same general results, then the methods are probably faulty and the discoveries usually are considered invalid.

Has any scientist ever assembled a living organism, even a single bacterium, from its component parts? No. Organisms assemble themselves. Organisms contain the instructtions within their DNA to carry out this assembly. So nobody has ever *proven* that life can be started from scratch using only chemical elements such as carbon, hydrogen, and oxygen. However, we might also ask: what *has* science demonstrated relative to the assembly of organisms and their components from chemical elements? The answer is: plenty, indeed enough to be of profound interest. Scientists have demonstrated very clearly all of the following:

(1) Molecules such as carbon dioxide, methane, sulfur dioxide, ammonia, etc., exist in very many places throughout the Solar System and throughout the universe.

(2) An extremely wide range of environmental conditions exist, again throughout the Solar System and throughout the universe. Some of these conditions, especially ones generating heat and certain sulfur compounds, promote chemical reactions in which small molecules are combined into larger ones.

(3) A whole variety of large molecules and vesicles that operate to put various compounds into compartments and direct the flow of materials in and out of those compartments will self-assemble when these compounds occur in environmental conditions similar to those we know occur widely in the Solar System and universe, including in volcanoes on Earth.

Thus little by little, research project by research project, the scientific community is showing the plausibility of life's complexity arising from relative simplicity, especially given a range of conditions under which potential components interact with one another. With almost each issue of widely respected scientific journals, the probability that life exists elsewhere in the universe, as well as the probability that it arose on Earth without supernatural intervention, increases.

Thus the science of life's origin is focused on process and materials, especially those that happen and occur widely in the known universe. Anyone with access to a reasonably large university library can also gain access to much, if not all of, the research on events surrounding the assembly of large molecules and cell-like vesicles. Are scientists likely to ever synthesize a working eukaryotic cell, one with a nucleus, mitochondria, and Golgi apparatus? Probably not. Are they likely to build a virus from scratch, a quasi-living entity that carries out virtually all life functions so long as it can infect another organism? Yes; in fact, scientists already have the tools to build viral DNA from scratch, and the likelihood of this happening is great enough so that Homeland Security is probably concerned about the know-how getting into the hands of biochemically sophisticated terrorists.

There remains, of course, the main player in this ID vs. Evolution debate, and that is the matter of a bacterium. The central issue is whether there is a God who breathed life into that first bacterium by organizing inorganic chemicals into known substances and then putting those substances into an interactive package that consumed energy and replicated its kind. This issue cannot be resolved because we cannot even design studies to determine whether God exists anywhere except inside the human brain. And the debate continues because scientists chip away, study by study, at the problem of self-assembly, and nothing they have discovered yet leads them to abandon the idea that life can indeed arise, and probably fairly easily (in evolutionary terms), whenever and wherever proper conditions exist. But once a planet has a bacterium, or its equivalent, then the problem of evolution becomes much less difficult, including the evolution of eukaryotic cells, and ultimately, several hundred million years later, thinking, conscious, beings like ourselves.

5. What is Creationism?

> *It is possible that in the very earliest traditions on which the Bible is based, the creation was indeed the work of a plurality of gods. The firmly monotheistic Biblical writers would carefully have eliminated such polytheism, but could not perhaps do anything with the firmly ingrained term "Elohim." It was too familiar to change.*
> —Isaac Asimov (*Asimov's Guide to the Bible*)

Creationism is an intellectual movement or set of beliefs crediting a supernatural power, that is "God," with creation of the universe. Creationism comes in many different forms, but those used in American culture wars are based on various interpretations of the Bible and focused on the origins of Earth and its living organisms, especially humans. If you ask a typical person on the street where the Earth came from, that person is likely to say "God made it" without giving much thought to how, or when, or other details of creation.

Among politically active creationists, however, there has been and still is a surprising diversity of views about these details. The most extreme creationists believe that the Earth was created in the not-so-distant past (in geological terms) in six days—literally, days as we know them—and according to the sequence outlined in Genesis. The least extreme creationists still credit God with all creation, including special creation of humans in a Garden of Eden, but bend a little bit on the time and mechanisms involved. No creationist, however, admits to any significant creative force at work on Earth beyond the hand of God, although "microevolution" such as occurs with the well-known manipulation of plants and animals under domestication is often accepted as an observable, therefore likely, event.

Creationism has a long and rich history, but most of that history is post-Darwinian. Prior to publication of *On The Origin of Species by Means of Natural Selection or the Preservation of Favored Races in the Struggle for Life* there was virtually no distinct intellectual movement known as creationism, at least in the modern political sense, because nobody, or at least a very politically-inconsequential few, questioned, at least in a very serious, effective, and public way, the idea that God had created this planet and all that lived upon it. So what Darwin actually did was legitimatize the discussion.

After 1859, it became a valid intellectual exercise to argue about how diversity among plants and animals was produced in nature and this validation quickly—at least in historical terms—extended to questions about human origins. So for the purposes of this essay, the term "creationism" should be considered a post-Darwinian phenomenon. In its most liberal definition, creationism allows all kinds of Earthly processes, including evolution, but gives ultimate credit for building Earth, and presumably the universe, and starting life, to God. In creationism's most conservative form, the story of Genesis is a literal fact.

In general, when people argue about creation and intelligent design, they are rarely well-informed about the intellectual history of ideas surrounding the origin of life. Instead, the argument tends to quickly deteriorate into one of whether or not God exists or, in more vitriolic form, whether one has religion (*our* religion) or not; at least that is the subtext. But creationism itself has a long and varied history, much of it highly political, although in this case the political battles often have been between individuals holding different pictures of the creation process itself, and especially the theological issues surrounding various viewpoints. And because nobody knows for sure where the universe came from, all explanations for our ultimate existence are at least somewhat conjectural. Given what we do know about our universe, however, scientists' explanations for the origin of stars, planets, and living organisms are not entirely conjectural, and indeed, the leading ones are supported by a great deal of tangible evidence.

The Internet, of course, has massive amounts of material on creationism, evolution, intelligent design, criticisms of all kinds of creationist and evolutionary literature, etc. One of the best sites is www.infidels.org, not so much because of what it says, but because of the links to so many other publications, including many books, columns, and reviews. Another interesting site is www.talkorigins.org. Of course the web also contains a lot of truly illiterate garbage, too, and not only about Earth's origins. The last time I did a Google® search using "creationism definition" as the key words, there were about 24 million hits; "creationism" produces about four million hits, and "creationism vs. evolution" a little over eight million. Clearly our fellow humans have a lot to say about where we came from and they don't mind saying it, regardless of whatever evidence might support, or invalidate, their visions of our origin. This online discussion, however, is perhaps best put into context provided by another Google® search term, namely the two to three million links that appear when you look for "creation myths" or "creation mythology."

The Genesis story is only one of these myths, and not necessarily the most dramatic or allegorical of them. If "creationism" were given equal time to evolution in the public schools, literally and truthfully, the multitude of various cultures' creation myths would make for exciting and dramatic times in the elementary classroom. In the American political dialog, for example, you never hear a demand that the Aztec story of creation be given equal time with "evolution." But if that happened, then fifth grade children might be taught—to quote the web site ancienthistory.about.com—"Coatlique got pregnant by an obsidian knife through which she produced her one legitimate litter of moon and stars. When she got pregnant, shamefully, a second time, she gave birth to the god of war, Huitzilopochtli, who murdered his siblings."

It's easy to envision some pretty interesting art to accompany this story in pre-teen school literature. Christianity may be a powerful political force in today's world, but in truth it is only one of over 4000 currently recognized religions and scientific support for the Genesis creation myth is no stronger than that for any of those others. To claim otherwise is to exhibit Christian bigotry and compound the illiteracy inherent in creationism.

The Biblical story is relatively open to various interpretation, too, especially with respect to time, and is even more open to interpretation when one considers life forms that we now know exist but are not specifically mentioned in Genesis. Thus there is a lot of wriggle room for people who want to use the Genesis story for various reasons and in various ways, and depending on how the story is interpreted, some fairly heavy theological issues surface. To illustrate some of the problems involved in understanding special creation, we might consider the group of parasites known as tapeworms.

It takes very little knowledge of zoology to realize that any answer to the question of *when* God made tapeworms—that is, before or after their hosts—leads inevitably to an interesting theological discussion because quite different post-creation events must occur to explain these parasites' continued existence, depending on when they were supposedly made. Genesis 1:20-25 deals with the world's fauna, so we could infer that tapeworms were included in the categories listed ("every living creature that moves"), or were simply included with, and within, the larger animals mentioned, such as birds and cattle.

All tapeworms are obligate parasites; they do not survive outside of their hosts, except as eggs passed in host feces. Therefore, if God made tapeworms *before* He made their hosts such as the birds and cattle specifically mentioned, then those worms must have been either free-living or something we would not recognize as tapeworms. If they were unrecognizable as tapeworms, then that means they were later changed into tapeworms by some mechanism not mentioned in the Bible, and because we're discussing creation instead of evolution, that mechanism has to involve a decision by God to transform an existing, free-living, worm (we suppose it was a worm) into a new kind of worm, this one a parasitic, segmented, hermaphroditic, egg-producing machine dependent on its host's defecation for survival as a species.

In other words, if we do not allow evolution to create a tapeworm from a free-living ancestor, then we must allow God to accomplish exactly the same thing as evolution evidently accomplished, although for some mysterious supernatural reason. If God created tapeworms anew *after* He created their hosts, however, and furthermore, created them in their present form, then He purposefully made a parasitic, segmented, hermaphroditic, egg-producing machine dependent not only on its bird or cattle host's defecation for survival, but also on the eating of tapeworm eggs (= eating of host feces) by various invertebrates such as beetles in which infective larvae could develop.

As if the timing of tapeworm origin were not enough of a theological problem, the *reason* why God made tapeworms compounds the difficulty of rationalizing their existence. It's difficult to seriously discuss why God made tapeworms because such a discussion quickly becomes an exercise in creativity, carrying with it a strong dose of smart-aleck cynicism. What was God thinking when He made these wonderful parasites? What was His intent? What purpose did God have for such a creation? But let's do the exercise because it's a fairly instructive one in terms of what we might call "creationism theory," although it involves an attempt to read the mind of God, an activity some religions consider blasphemous and probably most consider impossible.

Nevertheless, let's try to answer these questions, beginning with the idea of a nice tapeworm in the mind of God, remembering, of course, that we could do these exact same thought experiments with any of the 100,000 species of molluscs, the 400,000 species of beetles, the untold thousands of roundworm species, and just to include plants, poison ivy. I'll admit that reading God's mind is about as easy as reading your next door neighbor's mind. That is, it is virtually impossible. But, to explain the existence of tapeworms, we need to give it a try. Thus we might begin by asking the question: given everything we know about life on Earth, why should God have made a tapeworm?

Before we can address this question seriously, however, we need to understand that God didn't make just "a tapeworm;" no, God made hundreds if not thousands of species of tapeworms and put them into sharks, bony fish, amphibians, reptiles, birds, and mammals, including humans. And He made these various tapeworms highly diverse, structurally speaking, with numbers of testes ranging from one or two to dozens if not hundreds and uteri that could be sacs, or networks of tubes, or even containers that grow in place of the first uterus. But God evidently had the most fun with tapeworm "heads," producing many different kinds with suckers, crowns of hooks, tiny hooks on suckers, and glands. He also created tapeworms of a wide range of sizes, from tiny ones less than an inch long to veritable giants, many yards long. Some of the former He put into dogs and wolves, but the really big ones went into really big animals such as whales.

So what was God thinking when He made tapeworms? The first and most logical answer to this question is: God wanted some device for keeping some of his most intelligent, curious, insightful, and creative humans occupied for their entire lives. He knew, because He was God, that intelligent, curious, insightful, and creative humans come up with all kinds of blasphemous thoughts, and furthermore, are not always big fans of organized religion. So He probably needed a way to involve these minds in some activity that prevented their intelligence and creativity from being applied to other activities such as war, especially war conducted in His name.

We have some historical precedence for assuming that God made things to fool humans into harmless behaviors, perhaps the best ones being fossils, which keep lots of people occupied, for example, paleontologists with science, or bloggers with arguing about evolution and creation instead of killing one another or running for public office so they can do public damage because of their willful ignoreance. Tapeworms are better than fossils in this regard because they are alive; thus their complex life cycles and physiology compound the problems of understanding them and make them all the more attractive for intelligent people who ought to not be using their brains to build weapons of mass destruction.

So, having figured out what God was thinking when He made tapeworms, or at least coming up with a candidate answer consistent with what we believe God's mindset to be, we can extend His line of thought, in the process addressing His intent and purpose in a more general, theological, manner. It is probably a pretty good bet that God's intent and purpose for making tapeworms was the same as His intent and purpose for making all the rest of nature that we currently know about, or whose existence we can easily infer from what we do know, namely, as a source of truly great mystery and wonder.

Having designed humans, even though He'd still not actually built one, God realized what an enormously powerful device would be the brain He had all planned out, and He understood that such powerful information handling devices often took on lives of their own, or at least seemed to do so, thus producing an emergent property we now know as the mind. We can almost hear God saying to Himself: hmmm; if I build this thing like I've actually designed it, then it's going to need something to keep it occupied, and I mean *truly* occupied, with grand, unsolvable mysteries such as maybe why these brains exist at all. So obviously He thought right away of tapeworms.

Whereas tapeworms would work fine as divergence from war for a few highly intelligent and secular people, the average person would need much more personal challenges, for example, the problem of where people came from. Also, one usually needs a microscope to study tapeworms, but microscopes were not invented until long after the Garden of Eden was abandoned. So God, being God, recognized immediately that people were something that other people could easily observe without a microscope, and He also realized that this problem of where people came from could keep people occupied even when they had little or no idea what kind of evidence might be use to solve it. In other words, ignorance was no obstacle when people decided to get into an argument over where people came from.

What God didn't realize, therefore, was that instead of simply keeping people occupied this problem would allow those same people to rationalize war as one of the legitimate ways to address the very problem itself. Thus God came to observe that His creation was quite capable of behaving in unexpected ways, which some of these created beings called "free will," and furthermore was capable of convincing itself that God Himself was inspiring such behavior. We can imagine God sitting by His Heavenly picture window, looking out over Heaven, and wondering whether He should have stopped His work with beasts and their tapeworms instead of letting His creativity run rampant to the point of designing some really smart apes.

The mystery gets deeper when we presume that God made tapeworms and put them inside those animals that are mentioned in Genesis, for example, cattle and birds (in the Revised Standard Version), and didn't tell anyone, at least any of the people who ended up writing the Bible several thousand years later. So tapeworms inside birds and cattle could be interpreted as one more game of hide and seek, sort of like fossils. Our observations could then be consistent with some theological conclusions about God's personality, namely, that He's a creative, ingenious, and loving entity who likes to play hide and seek. Alternatively, we could consider Him a wrathful and jealous God trying to punish any creature on Earth stupid enough to not live a clean life by infecting that creature with a worm. We are not legitimately able, however, to consider tapeworms a plague thrown down by a wrathful God because tapeworms in general are not very dangerous and certainly not as capable of social disruption as say, locusts, which most people call grasshoppers.

So what people usually think about tapeworms, that is, that they are nasty and dangerous, is counter to what God knew about tapeworms, which was that in most cases they were pretty benign. In fact, tapeworms are generally so benign that in the vast and overwhelming majority of cases you can't tell whether an animal has one unless you study that animal's feces and find eggs, or kill the animal and cut it up to find the worm itself. In only a couple of instances can you otherwise determine that an animal might have a tapeworm, and with those infections you have to know where the animal has been and what it's been eating, and you also have to look at some feces or, if the tapeworm is a larva, perhaps do a CT (= CAT scan = X-ray computed tomography) scan of the animal's (human's) brain.

So God could easily have made tapeworms simply for His own pleasure. After all, tapeworms are truly amazing and intriguing organisms that have kept some humans (made in God's image) occupied for lifetimes. It's probably too blasphemous, and too speculative, to claim that God didn't really make cattle and birds and all the other "creeping things and beasts" anew, but simply copied them from another planet He'd created a long time ago in a galaxy far, far away, not realizing that in the time since He'd created that other planet, those parasites had evolved from free-living worms that had previously evolved from primitive agglomerations of cells that had previously arisen from some rich soup of organic molecules.

In this scenario, the tapeworms would be an accident resulting from God's laziness and ignorance, His plagiarism, so to speak, but because of our respect for God, we can't consider Him to be lazy or ignorant or a copycat regardless of the fact that we really don't know anything at all about Him. We may *believe* lots of stuff about God, but we really don't *know* anything, including why He made tapeworms. Given the size of the universe, we also don't know whether God might have made so many planets and populated them with all kinds of plants and animals that He simply lost track of what had happened on them as a result of free will and evolution and thus we now have tapeworms through no fault of anyone, especially God.

In this particular case, instead of laziness and ignorance, decidedly human traits, God might have been up to His eyebrows in administrative tasks and simply didn't have the time to check whether there were tapeworms in the beasts he was copying for Earth. This situation is so familiar to most humans, especially those who have ever held administrative positions, that we assume a Supreme Administrator could easily have found Himself in a similar circumstance, and with similar results. So this interpretation of the origin of tapeworms on Earth is consistent enough with the "made in God's image" that it seems almost plausible, or at least not particularly blasphemous. Furthermore, this last explanation allows evolution to occur on other planets, ones we imagine but don't actually know anything about, and such evolution is then outside the domain of American public school education, thus inconsequential, and not worthy of discussion.

As mentioned above, tapeworms are only one group of organisms that present us with a major philosophical problem, namely, why they exist. We could have chosen any of hundreds of thousands of known species, or even the estimated several million species that are yet to be discovered, from bacteria to tiny primates hiding away in the Amazonian jungles. If the only creation we were worried about was that of shelled amebas in the ocean, for example, few if any human beings would care anything at all about their source, or their history. But the only people likely to argue about creation of shelled amebas are strikingly similar to those who might argue about the origin of tapeworms, namely, a bunch of scholarly nerds, probably tenured university professors, with access to laboratories, microscopes, molecular sequencing machines, and computers. So the general rule is that creationism works best as a political weapon when applied only, or at least mainly, to humans, because most humans really don't care very much about the vast majority of other species on Earth, and if you're dubious about that claim, start asking some proverbial people on the street their feelings about mice and mosquitoes.

Creationism also works best as a political weapon when it's kept simple, and focused on God, people, and human behavior, instead of discussed seriously as a philosophical or theological matter in its fullest implications. As you probably suspect, I tried to do the latter when I asked why God made tapeworms, although I just as easily have chosen ticks, fleas, and cockleburs. These creations are certainly nothing special compared to the other millions of non-human species that share the planet with us; they are, however, pretty good examples of species that humans might easily consider useless, or explainable only by resorting to the old saw that "God's ways are so mysterious that we shouldn't try to explain why He made something, only admit that He did for some reason we can't fathom." Nevertheless, if we are to discuss Creation—with a capital "C"—seriously, then we also must ask why God made tapeworms, ticks, fleas, cockleburs, and poison ivy. In other words, we must engage in this rather blasphemous thought experiment, namely, trying to interpret God's intentions, or reading the mind of God, relative to worms and other seemingly useless and irritating species.

Ronald Numbers, an historian at the University of Wisconsin in Madison has produced what is perhaps the definitive history of creationist ideas, at least as they are held in the United States. This history is documented in detail in four books, the first three being: *The Creationists* (Knopf, New York), *Darwin Comes to America* (Harvard University Press, Cambridge, MA), and *Disseminating Darwinism: The Role of Place, Race, Religion, and Gender* (Cambridge University Press, Cambridge, UK). The latter book is an edited volume, with Numbers' co-editor being John Stenhouse. The fourth book—*God and Nature: Historical Essays on the Encounter between Christianity and Science*—also is an edited one, by David Lindberg and Numbers (University of California Press).

Although in any discussion of competing ideas, the participants always seem to have books that their opponents should have read, or which are called upon as authoritative ("Have you read the *Left Behind* series?" etc.), Numbers' books differ from those typically mentioned in one defining way: they are serious, scholarly, beautifully written ones about the history of ideas. Thus they are generally devoid of polemics except to the extent that straightforward narrative of events and behaviors reveals polemics. It is impossible, for example, to simply tell what some of these folks did and said—routinely a matter of public record—without also revealing them to have rather unsettling personalities and agendas, often, if not typically, heavily laced with ignorance about the natural world. The next two paragraphs are a distillation of Ronald Numbers' books mention above. I strongly encourage the reading of these volumes and their purchase for your personal library.

Creationism is not a very simple theological phenomenon. Throughout history, creationists have argued among themselves, for example, over the definition of "days" in English translations of Genesis. Some believe a "day" was indeed 24 hours long, as we now know it, or at least call it; others believe the "day" was some unspecified great length of time, with the term "day" being used metaphorically. Creationists also differ quite a bit on their interpretation of the Flood. Some believe the Earth is only a few thousand years old and that the Flood did indeed kill everything except Noah and his family, in the process creating the Grand Canyon. Others believe that Earth had a long history before the Flood, maybe even millions of years, in which the dinosaurs lived, but that God destroyed everything and started over, so that the Genesis story is actually a second creation, or maybe a third, or even the latest of several. The important thing to remember in reading through the history of creationist ideas is that people are not disagreeing over scientific observations. Instead, they're disagreeing over words in the Bible, and claiming mutually inconsistent interpretations of those words to be the absolute truth.

Creationists also disagree on what, exactly, God created. Some, in fact probably most who call themselves creationists, believe not only that the Earth is quite young (less than 10,000 years old), but that God created everything on it, especially humans. At the other extreme are those who believe that God created life, and perhaps endowed humans with a soul, but the rest of Earth's history, including biological evolution, happened more or less as scientists have described it in libraries full of published research.

For creationists, the so-called "higher taxa" have always been more of a problem than species or strains. Thus few would argue that agricultural researchers can produce a variety of food crops from a common ancestor, the descendents of *Brassica oleracea*, or wild mustard, being a prime example (broccoli, kale, cabbage, brussels sprouts, etc.) Those same creationists would argue, however, that the origin of larger groups such as antelopes (Artiodactyla), bears (Carnivora), mice (Rodentia), and whales (Cetacea) cannot be explained by natural selection, thus must be the handiwork of God.

The take-home message from this discussion of creationism is that humans are quite capable of constructing worlds of the mind, built only of beliefs, images, and ideas. Such worlds can have absolutely no relationship whatsoever to the real world of dirt, rocks, plants, animals, microbes, fungi, wind, and water, but we can live in our constructed worlds completely convinced they are the real ones. In other words, the intelligent designers are us.

6. What is Science?

> *The most productive scientists, installed in million-dollar laboratories, have no time to think about the big picture and see little profit in it . . . It is therefore not surprising to find physicists who do not know what a gene is, and biologists who guess that string theory has something do to with violins.*
> —E. O. Wilson (*Consilience*)

Science is a formalized study of nature, with the intent of discovering mechanisms that produce, maintain, and control natural phenomena. The term "natural phenomena" can include everything from the birth of galaxies to mating behavior of squid, or anything between any other two examples of the extreme and arcane. Not all of these natural phenomena are so strange, however; many of them are quite familiar, for example, infectious diseases (chicken pox, mumps, common cold, HIV), hurricanes, earthquakes, and the birth of human babies.

Some of these natural phenomena are of great interest to us because of personal, political, or financial reasons; others simply stir the curiosity or appeal to our aesthetic sense, e.g., shenanigans of backyard squirrels and a magnificent sunset. But regardless of our personal interest in a natural phenomenon, or lack thereof, some scientist has probably studied that same phenomenon with an eye toward explaining its origin, developmental processes, and the forces that control it. Science is therefore a way of knowing, but unlike other ways of knowing, science actually produces something, namely, technology, that can, with proper refinement or development, be used to control some of nature.

Science is not the same as technology, although science uses technology in its exploration of the universe from distant galactic clusters to molecular events deep within a cell. Conversely, people who seek to develop technology usually rely on scientific discoveries. Thus the two areas tend to complement and enable one another. Technology is to science as hammers, saws, and squares are to carpentry; you can't do the latter without using the former, but you can possess the former with no ability or intent to do the latter. Hammers, saws, and squares don't make a person an accomplished carpenter any more than a microscope makes a person an accomplished scientist.

Thus science is an activity in which *people* (scientists) use *devices* (technology) to explore the natural world. When that exploration produces something that solves a problem (defined by *humans*), then we are grateful and have a positive feeling about the science, a good example being the technology of vaccination against smallpox or polio. When the exploration produces a vision of the universe that we don't particular want to hear about, however, we sometimes have a negative feeling about the science involved, good examples being Galileo's explanations of planetary movements and, of course, the emergence of humans during the course of vertebrate evolution.

Thus science is easily, if not best, described by that tired old metaphor of the two edged sword: it gives you power but it can also hurt you. It can give you happiness, but it can make you angry and frustrated. In this encounter with the two-edged sword, it is well to remember the difference between science and technology, and to remember that both result from *human* activities. Humans have a way of producing phenomena that seem to take on lives of their own—large government agencies are an outstanding example—and science is certainly one of these although the world's total scientific enterprise is much larger, more diffuse, and less controllable than even the Social Security Administration or the Department of Defense. Science is so extensive and diffuse because, probably like religion, it is based on a fundamental human characteristic: curiosity, and most probably the curiosity derived from both consciousness and wonder.

We often describe non-human animals, especially our pets, as being curious, but we have no idea whether such apparent curiosity is based on wonder or whether other species are conscious in the same sense we are aware of ourselves and of the deeper meanings and symbolism in our environments. Many humans, especially in developed nations, will argue that their dogs and cats are "almost human" and in the worst case scenario people and pets come to look like one another, especially if both live long enough. Pets are not people, however, and just because we like to hang that baggage on them, our skewed view of their fundamental biology doesn't alter their genetic makeup very much no matter how successful they are at teaching us when to feed and pet them. I was discussing this matter one day with a colleague who used his girl friend's schnauzer as an example.

"This dog's lying there on the floor asleep and lets out a big fart," says my friend. "Then he suddenly wakes up and starts chasing his own fart and barking at it." Such behavior was cited as evidence that this particular pooch had so little self-awareness that it was not able to distinguish his own combination of methane and sulfur dioxide emissions from those possibly produced by a rival. I'm not sure that this particular schnauzer's behavior was typical of all dogs, but the case certainly illustrated clearly the extent to which the pet's owner and guest were people, aware of cause and effect, and willing to pass judgment on another species' traits.

If that schnauzer were a scientist, he would probably be wondering why that gas smelled so bad, and thinking way ahead to a whole lot of neurobiology experiments to trace the responses of certain kinds of nerve endings to chemical compounds diffusing through the atmosphere, all resulting in socially aggressive behavior toward an imagined intruder. Maybe in the back of this scientist's mind is a thought that perhaps such research might eventually lead to the ultimate aphrodisiac. This difference between the schnauzer's and the imaginary scientist's respective reactions to a natural phenolmenon illustrates clearly the roles of wonder and curiosity as stimulants to research.

Science sometimes gets a bum rap from a lot of people for being boring and difficult to understand, if not outright unpleasant. If you talk to enough people about science, you routinely hear the expression "I'm just not good at science," which could mean anything from "I'm just being polite by not telling you [a scientist] that I think science is evil and only really despicable people become scientists" to "I made a D in freshman biology and have never forgiven my teacher or adviser even though that happened 37 years ago." But the phrase "I'm just not *good* at science" often really means "I'm not the least bit *interested* in science unless, of course, it involves a matter that affects *me or my family right now*." A low-level nuclear waste facility, planned for your county, turns even grandmothers into amateur nuclear physicists. A childhood behavioral disorder diagnosis turns a young mother into a medical researcher as quickly as she can get to the computer and call up Google®. And any perceived threat to someone's religious beliefs can easily transform that person into a philosopher of science even though he or she may not be able to tell a Mississippian ostracod fossil from a Devonian foraminiferan.

Of course these three individuals are not *really* nuclear physicists, medical researchers, or philosophers of science respectively; they just think they are, and believe they are behaving in a manner consistent with the behavior of their models. And in truth they are behaving like scientists, although not nearly as effectively, and with only a tiny fraction of the resulting knowledge and understanding that a professsional scientist would obtain from similar efforts, aimed not at Google® but at the "primary literature," i.e. published scientific research papers. Furthermore, the local citizens' study is likely to be guided by personal or political goals: to stop the building of this nuclear waste facility, to "help" the physicians treat his or her child, and to change whatever is being taught in public schools.

The scientist's study, on the other hand, is most likely driven by curiosity or a desire to build, or maintain, a reputation because respect for one's opinion is the currency in science. If the nuclear waste facility is built because a real scientist has demonstrated, convincingly, that it poses no threat to people or wildlife, then that scientist has bragging rights. If the nuclear waste facility is not built because another scientist can demonstrate, convincingly, that its design is flawed and that the leakage will indeed be a public health hazard, then that scientist also has bragging rights. A year after the decision, the locals may be either sighing in relief or seething with bitter anger, but there's a very good chance the scientist is off somewhere else bragging about his/her role in the political process.

In the culture wars, it often seems as if the true battle is being fought over thought processes, over mindsets. The question—Do you believe in evolution?—asked by reporters of American presidential candidates during a so-called debate during 2007, serves to illustrate this kind of battle. "Believe" is a completely human trait that has little or nothing to do with "fact," except that belief typically drives actions. In other words, we can believe something whether that something is true or not, or even based in reality, then act on such a belief.

We might believe that we *will* win the PowerBall® lottery, for example, although such belief is based on infinitesimal odds, indeed, on odds so tiny that they approach zero. On the other hand, we also believe that we *can* win PowerBall, and in this case the belief is well-founded because of two reasons: first, however small, the odds are not in fact zero, and second, somebody actually does win periodically, demonstrating the inevitability of a winner that could just as easily have been one of us losers. If we truly understand what it means to "believe" something, then we also understand the difference between "will" and "can" and this understanding probably increases the ornery pleasure we get from buying a ticket and imagining what we would do with all that money, knowing all the time that such mental entertainment is worth far more than $1.00 regardless of the ultimate outcome of this week's drawing.

"Evolution," on the other hand, unlike PowerBall®, is a realm of scientific inquiry, a process that has been demonstrated time and time again to work, and a theory. One does not "believe" in realms of scientific inquiry, processes, and theories. Inquiry about process, using theory as a guide, provides a particular path toward understanding the process. In other words, to a scientist "do you believe in evolution?" is a pretty stupid question although to a politician, or a reporter looking for political sparks, it is a great question. This difference in quality of the same question residing in different people's minds is derived from the respective mindsets of reporters, politicians, and scientists. We believe that we understand politicians' and reporters' mindsets; the former wants to win and will do or say about anything it takes to influence voters, and the latter wants a good story, preferably by deadline. If our belief about politicians' and reporters' mindsets is somewhat correct, then their behavior is more or less consistent with "fact," i.e., with observation.

But it is almost impossible to understand the scientific mindset without also understanding the fundamental structure of scientific inquiry, "structure" being defined as the components of a system and interactions between such components. Scientific inquiry is built upon observations, ideas, and approaches to problem solving, all human endeavors. This structure—the presence of and interactions between observation, thoughts, and approaches—helps explain why science itself is a distinctly human trait, certainly as human as religion. Chimpanzees may explore their immediate environments, but their techniques and approaches are not at all what we would call "science."

Chimpanzees make observations, remember whatever they have seen, sometimes act on their experiences and memories, and can pass along learned behaviors, sometimes unintentionally. But they don't do science. Insofar as we know, they don't formulate hypotheses, don't do statistical analysis, don't use mathematics, don't publish in journals, and don't design experiments. Nor do chimpanzees assimilate observations and infer causality, developing real theories the way astronomers do. In all fairness to chimps however, they are a long way from what we would call "dumb" and they exhibit some exceedingly human-like traits such as emotions, quasi-politics, duplicity, and competition. But they don't do real science. Or, to phrase it more accurately, we have never observed them doing what we would call real science, or if we did, we didn't understand what the chimps were doing.

Science as it is typically taught in public schools, and as usually presented to the public, is mostly what we would call "proximal" science; i.e., it is based on questions that begin with the word "how." An example of such a question might be "How is sugar broken down into energy, water, and carbon dioxide?" This question is an important one, obviously, because it concerns fundamental metabolic processes that our bodies use continuously and without which there would certainly be no human life as we know it today. Similarly, one might ask "How is the sun's energy captured by living organisms?" This question is an important one, too, perhaps more important than the one about sugar metabolism, because if life could not capture the sun's radiant energy then there would be no life at all as we know it—no cereal grains, no olives, no potatoes, no domestic livestock, no chimpanzees, no tropical forests or coral reefs, etc. But both of these questions are ones addressing function, and both have been answered by the chemists who study reactions between molecules.

In the case of proximal ("how") questions, we can usually "prove" an answer, especially in those cases where we can provide ourselves with material to repeat experiments. We can grow corn seedlings by the millions if necessary, grind them up into cytoplasmic soup, and carry out chemical studies on their metabolic processes over and over again. We also can manipulate these corn plants genetically by skillful, indeed Darwinian, breeding and selection; thus we can eventually associate chemical properties with inherited variations. So we are able, by formulating insightful hypotheses, and doing laboratory studies on our abundant material, to demonstrate clearly that corn seedlings carry out a series of chemical reactions that lead to the capture of radiant energy. That is, we can *prove* something happens regularly and such proof usually satisfies a critical public.

The kind of science I have just described also is what Thomas Kuhn (see *The Structure of Scientific Revolutions*) would call "normal science;" that is, it addresses accepted questions, ones that the scientific community itself considers important and legitimate. We know plants capture the sun's energy, and we know that we depend on plants for food because of that ability to capture energy, so the processes by which this capture takes place are legitimate ones for serious study. Maybe if we know enough about those processes, we can improve upon them, or use them for some other purpose such as making ethyl alcohol to fuel our automobiles. Although we could also ask "how can corn plants get dates to the prom?" that question is not a legitimate one for a variety of reasons.

But before we ridicule someone who asks how corn plants get dates to the prom, we should remember that it has not been too many years since questions about genetic engineering were not legitimate ones, either. Many manipulations carried out routinely today would have been considered science fiction at best, if not outright fantasy, a generation ago. In addition, if you go to enough scientific lectures, there is an excellent chance you will hear someone use the phrase "date to the prom" in a metaphorical way to describe succsessful mating events between organisms that have never heard of a prom and are never likely to, either. Thus farmers plant corn in rectangular plots in part because the plants are wind-pollinated grasses and so square-ish fields maximize attendance at the prom, metaphorically speaking, although I'm not sure many of those same farmers, watching their own daughters heading out to the Senior Prom in a nearby town, would be comfortable with the full implications of the scientists' metaphor.

In contrast to asking how corn plants trap energy, if we ask "why" corn seedlings eventually produce corn seeds, however, we suddenly have a question that is unanswerable except in evolutionary terms. In other words, we have an ultimate question—one of origin instead of a proximal, or functional one. Public school and televised science do quite well with proximal questions but they fail miserably at ultimate questions. The reason they fail is because they do not address the nature of evidence, the process of inference from comparative observations, and the testing of historical hypotheses. These three subjects are fairly sophisticated ones for both school children and the general public. In the latter case, there is not much evidence that the general public even cares *why* corn plants produce seeds, caring only *whether* they produce seeds, and in what volume. Yet it is the "why" question that sustains our culture war over evolution.

When scientists seek answers to "why" questions, answers that do not involve supernatural forces—e.g. an Intelligent Designer—then those scientists seem to be encroaching on a realm of intellectual and spiritual life that historically has been territory controlled by the formal religious establishment. We all know, from our study of history, what happens when science confronts a formal religious establishment, especially one with considerable political power. In one of the most familiar examples from history, Galileo was sent before the Inquisition and imprisoned for championing the idea that the Earth revolved around the sun and not vice versa.

Although today that particular condemnation and imprisonment would seem outright ridiculous, especially for publishing material claiming the Earth orbits the sun, remember that in the 1500s the geocentric universe was a product of extreme ignorance. Thus the real issue in our current cultural war is not whether "evolution is a fact," but whether ignorance fuels a controversy with serious political overtones, especially in a heavily armed nation that is technology dependent and losing economic ground daily to more secular societies such as China.

Galileo's troubles resulted from an entrenched political system in which belief trumped evidence, indeed, in which belief was taken to be evidence. There is certainly nothing in the historical record that suggests contemporary individuals or even modern societies are immune to troubles arising from this same situation, namely, belief trumping, if not being taken as, evidence. Perhaps the best documented example of such trouble is the fate of New Orleans at the hands of Hurricane Katrina in 2005. Katrina was the most destructtive and costly such storm in United States history, and much of the destruction, estimated at over $80 billion worth, and the loss of nearly two thousand lives, can be attributed to erosion of coastal wetlands and inadequacy of protective levees. Scientists had warned for years that such disaster was inevitable, and explained why, such explanation being founded on clear understanding of coastal ecology and global meteorology (see the cover story of the October, 2001, issue of *Scientific American*).

Of course politicians ignored such warning, and just as of course, the reason was scientific illiteracy and inability to explain, to the tax-paying public, why massive expenditures would be necessary (for levee construction), why coastal development magnified the danger (loss of a natural buffer zone), and why the question was not *whether* a Katrina-level storm would hit, along with the attendant damage and loss of life, but *when* it would hit. Science, as a discipline, remember, focuses on evidence; in this case there was substantial evidence to support scientists' predictions, enough evidence, in fact, for the scientific community to consider New Orleans somewhat of an historical "experiment."

Historical experiments are rather like so-called "natural experiments" in the sense that humans watch while nature does something to test a prediction. In science, belief results from a negotiable interpretation of evidence. We believe that a statement about nature is true until it is demonstrated otherwise through the use of observation, hypothesis testing, and appropriate evidence, such evidence sometimes arising through observation of a "natural experiment." In the case of Katrina, the obvious null hypothesis to be tested was: There is no difference between tropical storms. In this case, nature performed in a way that provided appropriate evidence and that hypothesis was rejected.

The scientific description of this historical experiment is almost clinical—"That hypothesis was rejected."—but the human and economic aftermath of this decades-long experiment is an on-going sad lesson about the social impact of scientific illiteracy by people in major positions of power and responsibility. No candidate for public office should be expected to have scientific knowledge equal to that of a professional scientist. But every candidate for public office should be expected to understand why scientific evidence is important to society, why in the long term ideology never trumps nature, and why he or she should be articulate enough to explain, honestly, the same way a scientist would be forced to by anonymous peer reviewers, why belief about how nature behaves should never drive public policy when it is obvious that such belief is ill-founded and not supported by the observations.

7. What is Religion?

> *Archeologists have unearthed Neanderthal graves containing weapons, tools, and the bones of a sacrificed animal, all of which suggest some kind of belief in a future world that was similar to their own.*
>
> —Karen Armstrong (*A Short History of Myth*)

Religion is a human activity in which some superhuman, and supernatural, power is held in high esteem, given credit for creation of the universe as a minimum and typically for many other phenomena, in which behavioral norms, or at least ideals, are established on the basis of written documents (or their oral tradition equivalent) presumably emanating from that supernatural power, and in which belief not only trumps observation, but also routinely dictates how observations are interpreted. Depending on the religion, there could be several to many such supernatural powers, perhaps with various tasks and areas of responsibility partitioned out amongst them. The behavioral ideals are ones often deemed necessary for survival of a particular culture, and in the primeval past that viewpoint might well have been valid. If this opening definition sounds a little academic, remember that we are not talking about *your* religion as *you* practice it on Sunday; we are talking about *all* religions, insofar as that ambitious conversation is even possible.

Some sources suggest that humans have produced, and practiced, at least 200,000 different religions, although according to our quick-check source, the Internet, there are currently around 4,000 of them. A visitor from outer space, spending a few years on Earth, would quickly come to the conclusion that religion is characteristic of the human species, perhaps as defining as any structural feature. If such a visitor were a scientist, he, she, or it would likely proclaim that the dominant primate on this particular planet is quite capable of believing just about anything in the face of evidence—sometimes massive amounts thereof—contradicting that belief, provided the belief is tied to some supernatural power.

We don't know whether non-human animals are capable of believing in the same sense that we are. What we do know is that human beings not only are quite capable of "belief," but also equally capable of considering their belief to be "faith" routinely evolving into "truth." In the English language, "faith" is an exceedingly noble trait but "belief" can be pathological. Nevertheless, the two terms, and their meanings, are very close and perhaps intertwined, particularly in the realm of religion. This intertwining is neither good nor bad, but it is clearly a feature of virtually all religions, and it is the reason science and religion have so often been in conflict, especially over the noun "truth" and the derived adjective "true."

Nobody knows if there is a God, and in all honesty, regardless of what many of us *believe*, nobody *knows* where the universe came from, either. The culture wars are sustained, in part, because of this complete lack of real knowledge about the two defining entities in our conscious human lives—the reason for our existence, and the physical realm that allows such an existence. Because we don't know where the universe came from and why it and we are both here, we have the power to create such "knowledge" in our minds. That "knowledge" then becomes belief morphing into faith evolving into "truth."

We're not completely certain of the extent to which non-human species are aware of their existence and can question certain facets of it. Recent observations suggest that porpoises gossip about other porpoises, a sure sign, some would say, of self-awareness, and elephants have long been credited with amazing insight, perception, emotions, and self-awareness. Bonobos, or pygmy chimps, our closest relatives, are exceedingly aware of themselves, activities in their environment, and social situations. But we really don't know if elephants, porpoises, and bonobos wonder where they came from either as individuals or as kinds, nor can we determine whether chimpanzees—bonobos' closest relatives—are curious about where termites come from even as they eat these insects by the thousands. So religion that addresses these questions of origin and reason for existence must be seen as a product of self-awareness that goes well beyond that possessed, insofar as we know, by species other than humans. This self-awareness is actually consciousness, our defining mental trait, although again even that assertion must be hedged by declaring that we are orders of magnitude more self-aware than other species.

If we humans were not so aware of our own existence, as well as the existence of other species, then probably we wouldn't care so much about why we are here. I contend that even if we had absolute proof enough to satisfy any rational person that there was no God, people in general would believe in one anyway, probably to the extent of starting a fight over the particular form of their belief. That contention comes more from the historical record than from any special insight into human nature. We've been fighting, routinely to the death, over our various visions of God for centuries and nothing in your daily news suggests that conflict is going to end any time soon. In fact, our daily newscasts indicate rather strongly that religion not only is a deadly force in human affairs, but also is becoming more so, especially with the massive global traffic in highly portable and lethal weapons.

Would a scientist from outer space, visiting Earth five million years ago, predict on the basis of primate behavior that in 2011 groups of *Homo sapiens* would killing one another, using religious beliefs as an excuse regardless of the underlying reasons. The answer is "probably not." Nothing in the behavior of existing non-human primates suggests they kill one another over their respective ideas about God. Bonobos and chimpanzees are the best living examples of what non-human primates were probably like five million years ago and we have no evidence that members of one troop consider members of other troops infidels worthy only of death. These apes may kill one another fairly regularly, and even conduct their equivalent of war—troop against troop with deadly consequences—but we cannot determine whether such action is based on religious beliefs.

Members of different chimpanzee clans may smell different from one another, or they may have behavioral traits we interpret as a hostile reaction to perceived threat or competition. The phenomenon we humans know as "the other" is manifested widely in the animal kingdom, although distinctions between "us" and "them" usually is made on the basis of chemical cues, especially in the notorious cases of ants, for example, who immediately attack intruders, even other ants, who wander too near the nest.

For humans, religion stands second only to blood relationship as a factor in making "us" vs. "them" decisions. Islam, the Church of Latter Day Saints (Mormons), and the Catholic Church are prime examples of this phenomenon, but the stronger Christian fundamentalist and evangelical groups also can be exceedingly tribal in their behaviors toward non-members. Depending on the sect, apostasy can result in shunning or even a death penalty. For example, in February, 2006, Abdul Rahman, an Afghanistan citizen, was sentenced to death by the courts for converting to Christianity more than a decade earlier. The ensuing international outcry stopped the execution and Rahman is now living in Europe, but the Hanafi school of Islamic jurisprudence still mandates death for apostasy.

In the post-Darwinian United States, the word "evolution" routinely has been and still is applied as a label for infidels, and Third Millennium reporters' queries of American presidential hopefuls, as to whether they "believe in evolution," is clearly an attempt to distinguish candidates from one another on the basis of religion. With a Mormon declareing his candidacy in 2007, and again in 2011, and ultra-conservatives such Rick Perry, governor of Texas and Michelle Bachmann, representative from Minnesota, declaring in 2011, it's fairly obvious that religion can potentially influence decisions made by the leader of an extraordinarily powerful, heavily armed, and belligerent nation—the United States of America.

Such introduction of religion into politics was not necessary, of course, to demonstrate leadership qualities equivalent to the task of running a great nation, but religion and politics have been bedfellows probably for as long as there have been humans. Indeed, Americans got over a serious litmus test for religion as a leadership issue when we elected John F. Kennedy. So such a question—Do you believe in evolution?—is really, whether intended to be such or not, a question about scientific literacy, addressed to a wannabe leader of a nation extraordinarily dependent on science and technology. The United States has not lacked for scientifically illiterate presidents; but we may have entered a new and dangerous era when candidates must prove to a scientifically illiterate public that they are themselves scientifically illiterate *because* of religion.

Like virtually all aspects of human endeavor and biology, religion comes in an almost bewildering range of forms and its influence on people is equally as varied. The average human, however, typically knows very little about the full spectrum of religious beliefs, and probably has only an abbreviated sense of the formal belief system of his/her own religion. Thus any conflict over or between religious entities is one that is a highly simplified argument over right and wrong. It's never very clear in such conflicts whether the "right" and "wrong" are phenomena that are truly dangerous to society or simply perceptions of right, wrong, and danger.

One might ask, for example, whether acceptance of a little bit of secular modernism is truly dangerous to a backward, fundamentalist, society. The answer, of course, is "no" but such modernism has a way of threatening the existing power structure, so that a belief system can become inextricably entwined in the processes of governance. One need only to do a Google® search on "Saudi Arabia" to find compelling evidence that this assertion is correct. There is nothing backward about Saudi Arabia, but there is plenty that is fundamentalist, and certain relatively modern and secular practices that Americans take for granted are clearly perceived as threats to The Kingdom.

Are the Saudi's "wrong" in their attitudes, their behaveiors, and their reliance on Wahhabism as law? No, they are not "wrong;" they are just different from Americans. Yet in the realm of religion "different" is often, if not typically, synonymous with "wrong," even to the point of condemning the "different" to death. Again, Google® provides ready access to background for such an assertion. Use "Sunni-Shiite conflict" as your search term; alternatively, read any semi-reliable printed news source such as a large city newspaper or one of the mainstream news magazines (*Time, U.S. News and World Report,* or *Newsweek*). Any adult American citizen who was living in 2007 knows that members of those two Islamic sects were killing and abusing one another daily in Iraq—including blowing up one another's mosques—in what was clearly a civil war.

For more of a historical perspective, visit your local library and look up "Irish Catholic" in the online catalog. For deeper perspective, consult any fairly serious reference on religious history (Karen Armstrong's *The Battle for God* and Jonathon Riley-Smith's *The Crusades: A Short History* are outstanding ones). As a species, we have five digits, an opposable thumb, upright posture, a highly enlarged cerebral cortex, and a propensity to kill one another over our beliefs. Religion thus functions to define membership in a group, and depending on the nature of a group, outsiders may be ignored, tolerated, engaged in scholarly discussion, considered targets for conversion to membership, or killed.

Aside from its role as a source of violence and hatred, is religion also a force for good? The answer to this question is clearly "yes," and again one need only do a Google® search on "Catholic Social Services" to discover an enormous amount of charity and altruism sustained largely by religious belief, or at least by the strong belief in the basic character of a particular form of religion, such character then driving behavior. Is the institution we know as "Catholic Social Services" self-serving? Probably. Is there any missionary-type activity conducted by Christian denominations that is not, at least in part, self serving? Probably not.

On the other hand, there is some evidence, in this case from the realm of artificial intelligence, that suggests religion, and especially that part of it leading to missionary work and other kinds of altruism, is sustained by infective, self-replicating ideas. I quote an example from the writings of Douglas Hofstadter (*Metamagical Themas*), in which he outlines an idea system that can easily become self-perpetuating and infective:

System X
  Begin
  Statement (1): Blah.
  Statement (2): Blah, blah.
  Statement (3): Blah, blah, blah.
  .
  .
  Statement (n): Blah, blah, blah, ...
  Statement (n+1): It is your duty to convince others that System X is true.
  End.

The Hofstadter example starts to make sense in a religious, if not in a primate evolutionary, context when he gives us a system that reads:

System X
  Begin

Statement (1): Anyone who does not believe in System X will burn in Hell.

Statement (2): It is your duty to save others from suffering.

End

This last kind of system is clearly self-perpetuating and capable of spreading throughout a human population fairly quickly, although the particular version of it we now know as "Christianity" did struggle somewhat in its early stages. With the conversion of Roman Emperor Constantine in 312 A.D., however, this set of infective ideas called Christianity gained enormous power and became epidemic, in the process solidifying the belief, morphing into faith and eventually evolving, in the minds of those infected, into "truth," that a man named Jesus was indeed the Son of God.

All religions have creation myths. Again, a search using Google® reveals hundreds of such myths and provides summaries of most of them. The Internet is not exactly a reputable scholarly enterprise, of course, but in this case it does give one ready access to literature, references, and sources that can be confirmed, if necessary, by true anthropological or historical research. In the case of the so-called "creation-evolution controversy," the important thing to remember is that all religions have creation myths. Thus in essence, when we claim that the opening verses of Genesis are the "truth" because they are the word of God we actually are saying "my myth is correct but yours is wrong if it differs from mine."

Similarly, when we interpret Christian scriptures liberally, claiming that the Bible is a metaphorical outline of a creation carried out by the Christian image of God, then we are also saying "my myth is correct but yours is wrong." And when we claim that the Bible is the word of God in all matters, including creation, then we are saying "out of all those ~4,000 known religions, mine is the only one that is valid, that provides a true story of the relationship between God and humanity."

Is science a religion? I address this particular question because it has been raised in the public discourse on evolution. We might begin by asking whether science has creation myths. The answer is a qualified "yes," because although our explanations for the origin of the universe are tested repeatedly by scientific studies using exceedingly sophisticated technology, we really do not *know* where the universe came from in the same sense that we know that sugar is burned into carbon dioxide and water inside our bodies. So some might say that if we don't know for certain how the universe came into existence, then our explanation for this event is really a myth. The "yes" must be qualified, however, because the explanations are at least testable. If the Big Bang theory is correct, for example, then we should be able to detect certain kinds of background radiation in space, and indeed, that is what has been accomplished with technology assembled for the purpose. What nobody knows, or will ever know, is what the cosmos was like before the Big Bang. This particular lack of knowledge may help explain why physicists tend to be more religious than biologists.

Does science have unassailable "truths" that serve a tribal function? Perhaps. Natural selection comes fairly close to being one, and biology is unique among the sciences in having a central unifying theme, namely evolution. For example, all molecular biologists, plant geneticists, experts in tropical disease, and medical researchers are able, and the vast majority quite willing, to put their work into an evolutionary context in a manner analogous to a rabbi, bishop, mullah, and Presbyterian minister having a scholarly discussion about the power of faith. A diverse group of scientists can have an amiable conversation about evolutionary mechanisms, or they can have a heated argument about which mechanisms might be operating in a particular system, but they would never argue about the *fact* of evolution, only its *form*.

Does this acceptance of evolution as a central unifying theme by biologists make evolution a religion? That is a question whose answer probably would consume several chapters, if not an entire book, and still not get answered to everyone's satisfaction. Do we build churches where people come to worship Darwin? No; we build laboratories where evolutionary theory is tested. Do we have a book that plays the same role for scientists that the Bible plays for Christians? Not really, although an argument could be made that Darwin's *Origin of Species* performs that role and Stephen Jay Gould's *The Structure of Evolutionary Theory* certainly matches the Bible's size, complexity, and explanatory richness.

But evolution is a natural process and the source of its power to shape living organisms is pretty clearly understood, revealed by scientists using a rapidly increasing toolbox filled with sophisticated technology. Thus scientific faith is focused on methodology, not a supernatural entity; that is, if a scientist does his or her work correctly, then the results of that work should uncover a secret about how nature operates, albeit such secrets typically are small and their revelation doesn't have much impact on humanity as a whole. In fact, the impact on humanity of the vast bulk of research in evolutionary biology is not even discernable compared to an act of willful ignorance such as committed by George W. Bush et al. when they used the phrase "weapons of mass destructtion" to involve the United States in a decade(s?)-long war resulting in untold death and destruction.

8. What is a Conflict between Science and Religion?

> *If Lévi-Strauss is right, myths are constructed by a universal logic that, like language itself, is as characteristic for human beings as nest-building is for birds.*
>
> —Lewis Thomas (*Lives of a Cell*)

A conflict between science and religion is actually an argument over the proper evidence to use when deciding what kind of mechanisms operate in nature. Scientists contend that such mechanisms are natural and can be discovered (for example, attraction between molecules, gravity, genetic variation); creationists contend that although these mechanisms can be observed to operate, they are not adequate explanations for the complexity and wonder of our planet. This argument can devolve quickly into a struggle between humans for control of something—most notably the minds of fellow human beings—although it remains fundamentally a disagreement about evidence and interpretation of observations.

When the "something" to be controlled is the mind set of a population, then presumably, as a result, the behavior of that same population, especially the moral behavior, also is controlled. People with strong religious beliefs tend to maintain that moral behavior cannot be expected from a population without embedded religious influence and education, often of a particular kind. Agnostics, including many typical scientists, however, may view religion as a public health hazard, given the violence that has been perpetrated in the name of God over the past centuries, and indeed is being perpetrated currently in various parts of the world.

Conflicts between science and religion are the rough equivalent of arguments between apples and oranges over which kind of tree produces the correct fruit. Notice I've used the term "correct" instead of "better." If you're an orange and if your grove of trees has established some criteria by which fruit is judged to be correct, and if these criteria actually are observable, then the argument *among oranges* can be resolved using available evidence. In other words, the oranges can decide who is really an orange, and whether that particular fruit has certain orange-like properties, including behavior (ripening time, color, display of navel, etc.). Similarly, if the oranges have criteria for relative quality, then they also can decide which oranges are *best*, that is, whether some oranges exhibit the roundness, shape of the navel, uniformity of color, all in a particular way. This line of reasoning is the same one operating in the editorial offices of the magazine *Sports Illustrated* when models are selected for the annual swimsuit issue.

The arguments can be resolved, that is, until the grove decides that criteria need to be changed, and then the argument is likely to be over criteria instead of over fruit. But any resolution is internal, that is, applicable only to oranges and only to that particular grove, although fruit from related groves may be having similar conversations about similar criteria. Any discussion among oranges is a good analogy for competition or conflict between sects or denominations within a major branch of some monotheistic religion, a familiar example being the arguments over ordination of gay bishops in the Episcopal Church or lethal conflicts between Sunni and Shiite Muslims.

Apples from the orchard down the road have no respect for orange decisions and judgments, nor should they; whatever properties it takes to be an apple, or a good apple, have no counterparts among oranges. It's only when the buyers come by, and have a choice of either apples or oranges, that the two kinds of fruit end up in competition for buyers' desires. Oranges want buyers to be fed oranges, and apples want buyers to be fed apples, and both want buyers to believe that the fruit selected is a *correct* one, although neither apples nor oranges have any inkling of the buyers' internal psychological or cultural milieu.

Again, *correct* is the operative word, not *better*, because in this example I have constructed, as in the creation vs. evolution culture battles, *correct* deals with evidence as well as criteria, whereas *better* usually is a judgment call, albeit often based on observable criteria, but just as often polluted by desire and opinion, except in cases such as war and athletic competition. In war and athletic competition, both sides can, although they not always do, agree on the criteria by which one competitor is judged *better*—for example, score, occupation, obliteration of culture, economic dominance. In such cases, *better* is not necessarily a permanent attribute, and winners are not necessarily *better* by any criteria.

In the political realm, of course, any culture war is really a fight over people's thoughts and beliefs, tied closely to desire and opinion. Thus candidates for public office in the United States get asked: do you believe in evolution? As suggested by the chapters dealing with scientific literacy, this question skips the creation vs. evolution question altogether and probably should be phrased: are you religious enough to make your potential constituency comfortable, to make these particular voters feel confident you are one of them instead of an alien? The question might be phrased more honestly as: what kind of *evidence* can you present that allows us to judge whether you are religious enough to make your potential constituency comfortable?

As we saw during the 2008 American presidential campaign, in the political realm, even relatively flimsy evidence can be used to characterize a candidate's religious associations with real danger, an obvious example being Barack Obama's middle name (Hussein), supporting the blogosphere claim that he's a Muslim, therefore dangerous, regardless of the number of people named "Hussein" who have died in Iraq fighting for or in cooperation with the American army.

Thus in the political realm, you can be made to look like a bad apple without much supporting evidence, or with evidence that is simply fabricated and given credence by repetition.

Buyers in general, and especially those looking for fruit, are pretty gullible and non-critical; they're waiting, and even wanting, to be convinced of something that will relieve them of having to make a difficult choice. Furthermore, they tend to have little patience with evidence, and especially with the search for it, unless that evidence has something to do with a murder, preferably one on television, although certain mystery and thriller writers still sell millions of books. But with this little parable of apples and oranges, I've laid out most of the basic structure of the conflict between science and religion over the origins of life on earth, especially human life.

The take home message is that decisions about *better* and *correct* can be made on the basis of evidence, but only within a relatively restricted realm and only then in a defined sequence: we must answer "yes" or "no" to the question of membership in a group, and given a "yes" answer, quality then can be judged according to agreed-upon criteria (themselves subject to change). A "yes" answer to membership thus also is synonymous with *correct*. Only one important, indeed critical, element has yet to be considered, and that is the fundamental nature of this evidence that is central to the argument over which fruit is the *correct* one.

For the purposes of discussion, in this parable, let us decide that the oranges' evidence is primarily received wisdom, codified in written documents, and an internal logic. In other words, somewhere in the oranges' library is a book that tells them what a good orange is supposed to be, regardless of how many in the grove exhibit the requisite characteristics. Given the basic nature of oranges' evidence, there is no basis for making predictions that when tested could actually change the fundamental properties of these documents.

Furthermore, the logic is internal because it starts with the assumption that whatever is received is factually true: for example, the Bible is the word of God, Noah built an Ark, lepers were miraculously cured, etc. Because this evidence is an existing set of documents from which an internal logic is derived, it is entirely possible to define apples as oranges if apples will only accept the oranges' view of fruit and criteria not only for being an orange, but also for being a good orange, as spelled out in the documents. Such acceptance immediately turns apples into oranges, *in the minds of the other oranges*. When the oranges do research, the purpose of such investigation is to discover things that support what the documents already say, or validate the received wisdom, or clarify something that has heretofore been ambiguous, but always within the context of the documents.

This research of the oranges has the function, whether intended or not, of strengthening, increasing the longevity, and diversifying the use of those myths that serve as moral guidance. Like all literature, myths are powerful tools for embedding ideals into behavior, an excellent example being the parable of the good Samaritan (Luke 10:25–37). And, like all literature (as well as art, music, and science), the story is best understood in context, especially that established by historical research outside the realm of oranges' received wisdom. In other words, based on our knowledge of tribal sociology during the early Christian era, knowledge that could have been, and probably would have been, accumulated whether that happened to be the time of Jesus' life or not, to be a Samaritan was to be hated by the Jews, and vice versa.

Thus when the Samaritan stopped to help the robbed and beaten man, he was carrying out an act of mercy and exhibiting noble traits unexpected of someone who would have been considered inferior, perhaps because of religious beliefs or ethnic background. Thus he, the hated Samaritan, was exhibiting moral behavior superior to the priest and the Levite. From a modern scientist's perspective, the take-home lesson is that we humans are a single species and that our capacity for good or evil is an individual property, not necessarily related to age, gender, national origin, political affiliation, religion, ethnicity, or sexual orientation. That's how a scientist easily could, and probably would, interpret the good Samaritan story.

When apples do research, however, their evidence consists of observations about the natural world, and consideration of this evidence allows one to make testable predictions. For example, one could easily measure an apple's degree of redness then test a prediction about the probability of that apple being bought. Or depending on laboratory facilities available, a testable prediction could be made about degree of redness and fructose content. The evidence consists of considerable knowledge, obtained by observation, not only of the various kinds of apples, but also of pears, cherries, and roses, all of which are related to apples.

Apples thus know that there are so many kinds of fruit that choices as to which one is correct is simply a subjective judgment on the part of some predator, often involving a decision based on availability. Thus we see that the real issue in this conflict between science and religion is not so much whether one or the other has a correct answer to the question of where humans (oranges, apples) came from, and whether they are somehow correct, or better, compared to apes and monkeys (cherries, pears), perhaps being made in the image of God, but rather what kind of evidence is appropriate to the discussion.

Science and religion generally have been in conflict over the nature of evidence throughout recorded history, although the intensity of such conflict waxes and wanes. The three great monotheistic religions have typically ignored science, officially, although once in a while something happens to focus the attention of individuals with a religious agenda on matters of science. Science, or more properly scientists, rarely if ever focus much of their attention on religion, and they certainly don't spend serious amounts of time trying to discredit religious beliefs except, perhaps, among their friends at a local bar (although see Richard Dawkins' books, listed in the bibliography chapter). So a conflict between science and religion will never be resolved because it cannot be resolved, and it cannot be resolved because evidence brought to bear on the argument is not applicable to both sides of the leading question of our time, namely: what is a human being?

The only testable prediction both sides can make about this question is that humans will develop several different answers, all consistent with a set of underlying assumptions specific to the intellectual realm in which the question is considered. Scientists will answer that question one way because they are scientists, and priests may answer it another way because they are priests. Both answers contribute to our understanding of our place in the universe, but the scientists are correct in claiming that *Homo sapiens* not only is a descendent of non-human primates, but also probably interbred with other species of genus *Homo* and is the sole surviving representative of a diverse bunch of congeneric (= also in genus *Homo*) forms that walked the Earth over the past few million years.

A truly massive preponderance of scientific evidence, including everything from comparative anatomy, to embryological development, to fossils, to structure of our genes—that is, exactly the same kind of evidence used to reconstruct the history of hundreds of other species—tells us that we are descended from non-human primates and that we share common ancestry with living apes, especially chimpanzees, although there is some argument over which species of chimp is really our closest relative—the Common or Robust Chimpanzee, *Pan troglodytes*, or Pygmy Chimpanzee (bonobo), *Pan paniscus*. People are sentenced to death every year in the United States on evidence far flimsier than the evidence supporting scientists' assertions that we share a common evolutionary ancestry with chimps.

Nations also go to war routinely on far flimsier evidence of aggressive action on the part of perceived enemies, divorces are granted daily on far flimsier evidence of bad spousal behavior, and billions of dollars, nay hundreds of billions of dollars, are spent on sophisticated weapons systems with virtually no evidence whatsoever that such weapons systems function to make us healthier, more secure, more free, better educated, and more adept at meeting the challenges all of humanity faces during the Third Millennium than we would be otherwise. So conflicts between science and religion, especially in the United States, have little to do with evidence and everything to do with belief.

Any argument between science and religion, conducted currently in the United States of America, is actually a fight over two things: (1) what American citizens *should believe* about where American citizens came from, and (2) what candidates for public office *should profess to believe about where American citizens came from* if they are to garner votes from a large segment of our population. The first of these issues is fairly inconsequential and public debate over it actually is a fairly healthy activity because such dialog reminds us of the role that evidence plays, or at least should play, in our perception of the way nature operates.

In this case, primate evolution is just another instance of nature operating, so falls into the same general category of all other planetary processes, including ecological ones that we ignore at our long term peril. The issue of what candidates for public office should profess to believe, however, is a much more dangerous one, especially if such candidates actually believe what they must profess to get elected, if such belief is truly counter to the way nature operates, and if, once elected, these erstwhile candidates achieve positions of real power in which they act out their beliefs that are incompatible with nature's processes.

Generally people in leadership positions have quite a bit of power to convince large numbers of other people to think, believe, and act in particular ways. Citizens usually want their leaders to be authoritative, strong in their convictions, and articulate enough to incite some action to solve "problems." Thus people typically respond positively to leaders, especially those with certain characteristics: male sex, tall and well-proportioned physique, relatively deep voice, comfort in front of an audience, ability to modulate the voice to convey self-confidence and stir the public's interest, etc. John F. Kennedy was a prime example, although there have been others (for example, Barack Obama).

The 2008 Republican candidate for Vice President of the United States, Sarah Palin, also has these qualities (except gender), and if she could back them up with thoughtful, analytical, and cosmopolitan conversation, as well as a record of multi-lateral and cooperative problem solving, she'd probably have been headed for political stardom. Only the future will tell whether Sarah Palin's scientific ignorance, worn so proudly like some military medal, will keep her out of positions of real power, or whether our nation will continually get opportunities to watch her in action.

The same statements could be made about potential 2012 Republican Presidential candidates Rick Perry and Michelle Bachmann, Tea Party heroes in 2011. George W. Bush was somewhat of an enigma in that he possesses few if any of these physical characteristics we associate with strong leaders, yet proved himself capable of talking a whole nation into a disastrous and ill-conceived military adventure that shows no sign of abating and could easily continue for years, sucking up resources that should be applied to the nation's infrastructure. A case probably could be made that the Iraq war (2003 - ?) is actually a conflict between science and religion, that is, a fight in which the nature of evidence played a central, if not defining, role, with belief trumping reality.

Science and religion also are in conflict because scientific evidence can change with new observations, especially observations made possible by newly developed technology. Within a given religion, however, the fundamental nature of evidence rarely changes, given that it consists mainly of received wisdom. This distinction is not absolute, nor is it free from debate. In science, for example, certain physical and chemical laws are as unchanging as assumptions about the Godliness of Jesus Christ. And in religion, sects often differ in their interpretation of scriptures, or change their interpretations over time, the ongoing scholarship and discussion surrounding the Koran, Bible, and Talmud being prime examples. We have ministers, priests, and mullahs who regularly use scriptures in various ways to address contemporary issues, and each such use has the potential to be considered a modification of evidence. But the fundamental distinction between natural and supernatural always applies: science deals with the former, religion deals with the latter, and they come into conflict whenever one crosses into the other's realm. The conflict then, inevitably, is over evidence.

Astrophysics and particle physics are two areas of science in which developing technology is having a powerful effect on evidence, the very nature of that evidence, and the ways that scientists develop hypotheses and conduct studies. Scientists in these two disciplines have always asked, then tried to answer, the question: what is the fundamental nature of the universe? For example, sophisticated mathematical techniques, combined with equally sophisticated technology, allow such scientists to test predictions about the kind of background radiation left over from the Big Bang (assuming there was a Big Bang). Equipment such as the Large Hadron Collider beneath the Jura Mountains in Switzerland allows particle physicists to address questions about the fundamental nature of matter and energy.

But the average citizen, even in developed nations, is generally not able, and rarely willing, to discuss these areas of science except in a most cursory manner using information gleaned from news sources. Yet the astrophysicists and particle physicists probably have more to say about all creation than do the evolutionary biologists.

Evidence surrounding religious belief also changes as a result of research, but again, as in the case of astrophysics, the average citizen is relatively clueless about the fundamental nature of such research and the actual evidence itself. Few of us read Greek, Latin, and Hebrew, few of us can interpret ancient art as a reflection of the society in which it was produced, and few of us have access to archeological sites or data derived from them. We let the scholars do all this research, discovery, and interpretation for us, then soak up the summaries, especially if written in an engaging way (see the references, writings by Karen Armstrong).

Nevertheless, when you attend one of America's relatively main-stream and intellectually secure churches, for example, a reasonably well-endowed Presbyterian one in a university town, you are likely to hear a sermon supported by research. Such sermons tend to put Biblical lessons into a historical context, as mentioned above in our discussion of the good Samaritan. In other words, you're getting educated about the human condition, encouraged to be a noble example of our species, and given historical reasons for wanting to do so. In this particular situation, you are not being told you're going to Hell for all kinds of affronts to God regardless of how inconsequential those affronts may be to humankind.

The take-home lesson from this discussion is one about evidence and observation of the natural world. Galileo's interactions with the Catholic church, and especially the Inquisition, are instructive in this regard (see Dava Sobel's *Galileo's Daughter*.) The evidence for a heliocentric solar system was strong, but the political and religious implications of that evidence were equally as strong. Galileo's eventual acquiescence to the church is the equivalent of modern scientists declaring that their conclusions about climate change were false simply in order to avoid censure, after having demonstrated conclusively that the Earth's climate is changing, and fairly rapidly.

In the opening years of the Third Millennium, that is, prior to January 20, 2009, such censure could easily involve loss of a government job or an official government policy of silence. At least such scientists would not risk being sent to Guantanamo Bay, the modern version of Galileo's prison cell from which he wrote letters to his daughter. They would, however, have risked being ignored, and officially discredited by being officially ignored. And in a highly technological society, embedded in a global economy dependent on global resources, such discrediting is a reflection of scientific illiteracy in high places, an illiteracy that is of danger to the society that tolerates, even encourages, it.

9. Why are Science and Religion in Conflict?

Events, which are the arguments of God, are stronger than words, which are the arguments of men.
—Albert J. Beveridge (*in a speech to the Boston Middlesex Club, April 27, 1898*)

It's not entirely clear that science and religion actually are in conflict. It is abundantly clear, however, that *people* have arguments, often heated, and sometimes lethal, over their beliefs both within the realm of religion and between science and religion. So "conflicts between science and religion" clearly are intellectual conflicts between people holding different beliefs, usually about the natural world and in many cases about a supernatural world. So in the context of this chapter, "science" and "religion" actually refer collectively to groups of people.

In general, religion attributes at least some of the properties of the universe to supernatural powers, whereas science concedes nothing to supernatural forces or beings. In the final analysis, it is this difference in views of the natural world that lead to different interpretations of observations about our planet, the Solar System, the Milky Way, and the universe beyond our galaxy. A literal reading of Genesis is an example of one such interpretation of observations about our planet and its occupants; a modern evolutionary explanation of our existence, based on research in geology, comparative anatomy, and molecular biology, is another such interpretation. Obviously these two interpretations differ, regardless of the fact that they both are based on the same natural phenomenon: the existence of Earth and all its inhabitants.

These examples illustrate the fundamental nature of conflict between science and religion, namely, an argument over causality and origin of the natural world. But this conflict also deals with purpose, morality, fate, governance, and behavior, especially, it often turns out, sexual behavior. In other words, people use their different beliefs about the origin of our planet and of humans to justify or support rules of conduct, including intellectual conduct (worship, reading choices, etc.), as well as intimate conduct within the privacy of one's own home or apartment (for example, use of contraceptives).

Thus science and religion tend to differ significantly in their explanations of *why* we are here on Earth, indeed why Earth itself exists, *what* our obligations are to each other and to supernatural beings, *how* we should fulfill those obligetions, if any, *when* we should apply our interpretations and explanations, and *where* we should profess our beliefs. At the extreme ends of our intellectual spectrum, religious fundamentalists believe that the *what, why, how, when,* and *where* should be in direct compliance with God's law or word, whereas off-the-scale secular humanists believe that all such decisions are well within a human being's rights and privileges resulting from having been born a human. These two beliefs are obviously in conflict, and because the conflict concerns *belief*, it is never going to be resolved.

The conflict also concerns evidence, especially the nature of evidence to support interpretations of our observations about the universe. History abounds with examples of arguments over evidence, arguments that are often settled, in historical terms, by observations made using new technology. An easily understood example of this role played by time and technology concerns the geocentric universe. In the opening years of the Third Millennium, nobody anywhere, at least insofar as we know, argues that the sun and stars orbit the Earth, any more than they argue that the Earth is flat and if you sail far enough toward the horizon you'll fall off the edge into a pit of demons (except, maybe, in Kansas, where the Earth is, at least to the human eye, often quite flat).

Nor in the opening years of the Second Millennium would there have been such an argument, because nobody was around to argue, at least very convincingly, that the Earth orbited the sun and our sun was part of a large collection of stars called the Milky Way. Everyone also believed, deep in their hearts if they even thought about it, that you'd better stay close to shore in your boat or risk falling off the edge of Earth into that hellish pit. Well, maybe the Norse were not so easily convinced, or didn't pay much attention to their geography lessons, but early Second Millennium Europeans in general lived on a flat Earth in a geocentric universe.

Post-Galilean science makes these Medieval European beliefs seem naïve, but remember, the evidence to support such beliefs was derived from Scripture and the words that delivered this evidence to the masses were those of Roman Catholic Church authorities. Renaissance and post-Renaissance science, however, steadily eroded the power of religion to explain natural phenomena, and like science or scientific-type activities throughout prior centuries, science was the source not only of technology, but also of ideas and concepts.

When technology was economically important, providing people with tools, weapons, construction equipment, etc., then it was heartily embraced, as it is today. The manufacture of steel is perhaps the best example of beneficial technology arising from pure scientific-type exploration. In the case of metal chemistry, exploration continued from at least the first millennium B.C. up to the present day, although purely scientific metallurgy, conducted using testable hypotheses instead of trial and error, is certainly post-Renaissance. But throughout the centuries, to our knowledge nobody has ever questioned the rationale, the discoveries, or the collective investment humans have made in research on ways to combine iron with carbon and other elements. Steel is just too valuable to stir up much controversy.

Ideas are a different matter altogether. People may use technology to produce dangerous items (poison gasses, nuclear weapons, heavy artillery), but ideas can acquire power of their own and move quickly through cultures, thus threatening political structures that in turn are sustained by competing ideas. The conflict between science and religion then becomes an argument over the validity of what people, especially those in positions of leadership, say about the fundamental nature of our existence.

Religious authorities often have held inordinate amounts of power regardless of the societies in which they are found. Middle Eastern mullahs are prime examples of such individuals, and their exertion of power is a familiar story from our daily newspapers. Such power stems entirely from religion and with the willingness of people not only to believe that religious texts are derived from God, but also to behave according to the interpretations of those texts by contemporary human beings who, for a variety of reasons, end up in leadership positions. Put simply, when science challenges literal statements of religions texts, the legitimacy of power held by people whose role in life is dependent on those same texts also is questioned.

The reasons stated above are the main ones for conflict between science and religion. In recent years, particularly with the rise of the intelligent design movement and active attacks on Darwinism by people with academic credentials (William Dembski, PhDs in mathematics and philosophy; Michael Behe, PhD in biochemistry), the public also has seen what appears to be a weakening in the status of evolution as the central unifying theme of biology. Nothing could be further from the truth; if anything, books like those by Dembski and Behe (*No Free Lunch: Why Specified Complexity Cannot be Purchased Without Intelligence,* and *Darwin's Black Box: The Biochemical Challenge to Evolution,* respectively) have awakened the scientific community, especially those individuals within it who are active in education, to the political and economic consequences of scientific illiteracy. And regardless of how sophisticated these books seem on the surface, and how solid their authors' credentials seem to be, they do indeed promote scientific illiteracy.

The consequences of scientific illiteracy are substantial, especially to a nation such as the United States of America with its enormous storehouse of nuclear weapons, its heavy dependence on technology, and a rapidly evolving population that is becoming noticeably more ethnically diverse and equally noticeably less well educated. I am not claiming, as some will read into the last sentence, that ethnic minorities are among the less well educated. I am claiming, and this claim is well supported both in the scholarly literature and in widely available data such as performance on standardized tests (see Internet sites in the references chapter), that our existing public school systems are quite ill-equipped to handle the demographic and cultural changes occurring at present in our nation, and that predictably, our ability to produce a skilled work force is being eroded by that failure. The scientist's mind set seems to work just fine as a teaching device when all we're discussing is use of some substrate by bacteria; such use avoids, however, the pressing question of human biology in a deteriorating environment, thus allowing the conflict between those holding various view of nature to continue unabated.

As indicated in earlier chapters, science and religion are two of several, if not many, "ways of knowing." What does this phrase "ways of knowing" mean? It means a human's mechanism for acquiring information and either assimilating that information into an existing picture of his/her universe, or using that information to build such a picture. The key words in this definition, of course, are "information," "assimilation," and "build." The information can be anything, and I do mean virtually anything, from the specifications for a space vehicle to your lover's cell phone number. If there is any human construct that is close to infinity, it's the contents and implications of this word "information." Some data show that the average American receives about 25,000 messages a day, all of it "information." This information is in addition to whatever is observed about one's environment that enables a person to walk, drive a car, ride a bicycle, etc. All organisms live in information-rich environments, but not all organisms have humans' capacity for both generating information and considering, as information, something another species would either ignore or could not detect.

"Assimilation" means to incorporate something, in this case information, into an existing entity. When immigrants get assimilated, they are incorporated into an existing society and become functioning members of that society, adopting some of the traditions, engaging in business typical of the society, and conversely, however, also adding some of their own ideas, literature, and beliefs to their new society. This analogy holds for raw information. When a message enters our brain, we have the option of allowing it to stay, take up residence, and become part of our mental processes, that is, be assimilated. Maybe I should modify that statement a little bit by saying we *believe* we have the option of allowing information to immigrate into our minds. Some messages have properties that make them downright infective. Often such communications strike deeply into our subconscious, hitting a spot that usually must be kept quiet by rational thought, a good example being our natural fear of "the other" (people who are obviously different from us).

This "infective" property of some messages is a strong contributor to political conflict between science and religion because these messages become so deeply embedded in our beliefs that they evolve into fact, at least in our minds. Then we act as if we're fighting over real, physical, important, phenomena, instead of just ideas. Our species has a long history of merging ideas with real, physical, events (as mentioned elsewhere in this book), so it should come as no surprise that we do it with infective messages.

Our example of "the other" is an excellent one to illustrate this point. Animals in general react negatively to outsiders. Ants kill other ants that don't bear requisite chemical signals on their bodies, backyard birds claim territories and defend them against intruders, and humans almost invariably avoid contact with other humans who differ in skin color, language, sexual orientation, general body form, and finally, of course, religious beliefs. Tolerance toward others is a learned trait, not an inherited one, and in fact tolerance must be learned so well that it overrides our natural—that is, genetically endowed—tendency to reject "the other."

So it should come as no surprise that people in positions of power and influence can define "the other" and many, if not most, of us will accept, then act upon, that definition. "The other" has been defined, and is being defined daily, if not hourly, by someone in the United States, as: homo-sexuals, pro-choicers, blacks, Hispanics, Democrats, immigrants, liberals, socialists, Communists, tax-and-spenders, and, of course, evolutionary biologists. Paul Fussell explains a powerful social role for definitions by citing World War II propaganda and control of information to support our nation's definition of Japanese as "the other," if not down-right non-human (see Fussell's *Wartime: Understanding and Behavior in the Second World War*). A number of other authors also have addressed this kind of group behavior, namely, that of dehumanizing others, often as a prelude to violence (see Montegue and Matson, *The Dehumanization of Man*).

This intellectual tribalism, involving a significant amount of such "other defining," is at the heart of any political conflict between science and religion, and especially when that conflict involves evolution. To a deeply religious person, obsessed with the belief that humans were made by God in His image, some scientist's assertion that we descended from ape-like ancestors is tantamount to blasphemy. To a scientist fully aware of the fossil record, the molecular evidence, and the behavioral traits that together define us as a naked ape, the belief that humans were made by God in

His image is tantamount to willful, if not outright dangerous, ignorance, especially if used as a basis for social policy and international relations. Each side defines the other as "the other."

Attempts to reconcile the two viewpoints are never very successful; see Robert John Russell's book *Cosmology: from Alpha to Omega* for an outstanding example of such efforts. Russell tries to make the case that subatomic phenomena are the mechanism God used to produce a universe in which formation of Earth was a foregone conclusion, and upon which evolution of *Homo sapiens* followed naturally in accordance with God's rules (= quantum mechanics) of the universe. "Natural theology" is a term for the idea that all of nature, including evolution, operates as revealed by scientists, but such operation is a result of God's action(s), either past or present. Russell simply pushes natural theology down to the subatomic level; he finishes *Cosmology,* however, with the assertion that the Second Coming and Rapture are pre-ordained by the laws of quantum mechanics, a stretch, to be sure, but one that cannot be denied by science because it cannot be studied by scientific methods. Inability to deny something does not make that something true; it simply makes it incapable of being studied in a scientific manner, that is, incapable of being shown to be either generally true or patently false.

Are there some areas of human endeavor that cannot be studied scientifically but which contribute immeasurably to our human experience? Of course; religion is certainly one of these areas, but so are art, music, and literature. However, we know where art, music, and literature come from: they come from people. Individual members of the species *Homo sapiens* make art, music, and literature; and so, although we may not like some of these products, or agree that they should be distributed widely—think rap lyrics sprinkled with four-letter language—we nevertheless understand their origin and rarely if ever argue about where they came from.

People themselves are a different matter, as are birds, lizards, oysters, bumblebees, mosquitoes, and flu viruses.

Those who hold conservative religious views consider these natural phenomena to have been created by God, although the timing and sequence may vary depending on the levels of conservatism. Those who seek scientific explanations for natural phenomena consider it plausible for all these things to have arisen, ultimately and eventually, if not inevitably, from abiotic soup. At the heart of this disagreement is the concept of *purpose*.

Routinely in university student writings I encounter this word—*purpose*—as applied to some structure or function. Elephants evolved a trunk, for example, because they *needed* that trunk to survive; it fulfilled a *purpose*. Scientists rarely use this term *purpose*, however, unless they are talking about humans and human artifacts. Instead, a scientist would describe some structure as having a *function*. The elephant's trunk, a product of evolution, now performs many functions, some of which may not have been performed by just a long nose. The distinction between *function* and *purpose* may seem like a minor, even trivial, one to many people, but the distinction is real and important.

Purpose is a word closely associated with other human traits such as goals, decisions, desires, plans, etc.; *function*, however, is a very neutral word. Organisms evolve functions as a result of having particular structures, or vice versa; human organisms carry out functions for a particular purpose and acquire, often by choice or creative efforts, structures to carry out those functions. Your hand *functions* to pick up a pen; your mind lends *purpose* to the act, using a structure (pen) invented and designed by humans, and your mind anticipates the effects of your writing on a target audience, even if that audience is only you. When this distinction between function and purpose is ignored, you get scientific illiteracy, and with it, a conflict between people holding different beliefs in the realms of science and religion.

Furthermore, the concept of *purpose* is extended and applied, at least in Christianity, to human beings themselves: we were produced, so the Bible claims, for the *purpose* of glorifying God. We don't always *function* to glorify anything or anybody, especially a God, although such failure can be interpreted within a religious context, namely as sin. But as stated above, to a scientist, there is no purpose to any species' existence, only a function, and that neutrality applies to *Homo sapiens* as well as coyotes, jellyfish, and dandelions.

Why we are here, on Earth, a single planet orbiting a medium-sized star in one of billions of galaxies in the universe, is one of, if not the, most enduring question(s) in human history. Of course aside from our local moon, we can't go anywhere else except vicariously through use of technology such as satellites and telescopes, so we don't really know whether we are alone in the universe or not. But we can easily *believe* we are alone, convince ourselves we are unique in the eyes of a God, created by that God, and act according to those beliefs. This act of mind over past acts of matter works very well at the individual level. In the realms of social policy and international relations, however, when belief trumps evidence, failure and disaster commonly soon follow.

In the end, science and religion are in conflict when religion intrudes into the realm of scientific knowledge, including scientific explanations, for natural phenomena. One person's supernatural explanation for a natural phenomenon will always, so long as there are humans on Earth, produce a small social conflict as a minimum, and potentially a large political conflict, as demonstrated repeatedly throughout history. Science and religion also are in conflict when the opposite situation occurs, namely science intrudes into a realm of knowledge traditionally owned by religion, a good example being the definition of "human being."

In many ways we are supernatural, or at least behave as such, creating stories and worlds that exist only in our minds then going to live in such places, although the most hard-core scientists would claim that the mind is a natural phenomenon just like a brain. But science has much less tangible explanation for the fundamental nature of a mind than it does, for example, for the metabolism of glucose by nerve cells. Thus the mind and all its products remain firmly located within the realm of religion, or at least within that realm of which religion is an essential and prominent part. History tells us that intrusion into foreign realms routinely produces conflict; thus it is a given that such conflict should occur between the realms of science and religion, and that efforts to resolve such conflict turn out to be far more educational than they are effective.

10. What is Evolution?

> *Man still bears in his bodily frame the indelible stamp of his lowly origin.*
> —Charles Darwin (*The Descent of Man*)

Evolution is the irreversible change, over time, of any entity, typically a kind instead of an individual, and usually in response to changing environmental conditions. In a biological context, this change results from differential survival of variants, although the variants must have inheritable characteristics that define their particular variation. Biological evolution, therefore, can apply only to species and populations, not to individuals.

But as typically occurs in the use of language, especially with words that have metaphorical power, the term "evolution" has been applied to about any kind of identifiable entity with an extended life, from species of birds, snakes, and worms to nations, religions, and government agencies. We speak of "evolving roles" played by armies, for example, as easily as some of us speak of evolving viruses such as HIV or evolving primates such as *Homo sapiens*. This general definition—irreversible change—is a rather simple one. To scientists, the real issues in evolution are not the fact of change, but the why and how of change. Thus people who study biological evolution rarely if ever ask whether it has occurred, or is occurring now, but why it's occurring and by what mechanisms. People who study irreversible changes in business and government wonder the same thing, that is, why and how, instead of whether.

Before we get too far into a discussion of evolution, it's probably necessary to explain this scientific interest in mechanisms. By the term "evolutionary mechanism" I mean a set of events in which genetic differences between members of a species are produced and sustained. At the outset, we should admit that individual members of a species do vary genetically, sometimes greatly. Were this not the case we would have no "breeds" of dogs or cats, no human ethnic diversity, and no variety of wine grapes derived from various stocks of *Vitis vitifera*. Mutation is the fundamental source of variants, and by *mutation* scientists mean a mistake in the copying of DNA.

In order for such mistakes to be heritable, they must occur during the making of gametes (sperm and eggs). Mistakes can also occur during the making of gametes but after the copying of DNA; in this case the mutations are called "chromosomal." There are a number of human conditions known to be caused by chromosomal mutations, Down syndrome being a fairly common example. Mutation thus produces genetic diversity, but the variants that contribute to that diversity get eliminated, sustained, or spread through a population by various means.

People who claim that evolution cannot explain the presence of humans on Earth are claiming, of course, that there is no fact of evolution, which in their minds means that the why and how questions are moot. Historically speaking, much of the objection to teaching of evolution arises from this concern about *human* evolution. If scientists focused their attention only on ameba shells or obscure orchids deep in the tropics, there probably would be no political efforts to eliminate evolution from public school curricula or dilute it with creationist pseudo-science.

Thus we are led to believe, by creationists, that the principles of biology that apply to agricultural crops, butterflies, and disease-causing organisms do not apply to people. This belief is why attorneys and elected officials in general are not in charge of biology class. The philosophical morass that one conjures up when excluding humans from one of the fundamental principles of biology is very large and complex indeed. I may drag you into that intellectual black hole later, but for now, let's keep the issues, and the ideas, relatively simple. I'm also going to lead you through some of the essential components of "evolution," meaning the terminology.

Throughout history various creationist writers have said, in essence, that *microevolution* could be accepted but *macroevolution* could not, or would not. So any culture warrior needs to distinguish between these two words. Microevolution refers to small changes ("micro" = "small"); macroevolution refers to large changes ("macro" = "large"). If I'm a researcher trying to produce a strain of wheat resistant to a particular fungus through selective breeding, then I'm applying traditional microevolutionary mechanisms to solve a problem of social and economic importance. The change from susceptible to resistant can be a small one, genetically and certainly structurally, but the change in agricultural output could be very large, depending on the particular wheat farming situation. People have carried out microevolution so often, for so long, and on so many different organisms from orchids to pigeons, that we have shown ourselves perfectly capable of making it happen. Indeed, Darwin used such history of agricultural research to support his thesis. But nobody has changed wheat into corn or apples by selective breeding.

The next time you're in the grocery store, however, stop by the produce section for a study of *Brassica oleracea*, the wild mustard of Western Europe, which grows naturally on limestone cliffs because of its tolerance for salt and its intolerance of other plants. Its life cycle is biannual; during the first year it stores nutrients in its leaves and in the second year it uses those nutrients to build a tall flowering stalk. This first year nutrient and water storage makes the plant desirable as a vegetable, although the fact that it is not poisonous helps its popularity. *Brassica oleracea* has been cultivated for a great long time, perhaps thousands of years, and it was grown in gardens by the Greeks and Romans two thousand years ago. Today, you know it as kale, broccoli, collard greens, Brussels sprouts, cabbage, cauliflower, and kohlrabi. So the question of whether can you make broccoli out of mustard has already been answered in the affirmative, and again, Darwin himself used this kind of information to support his ideas.

In some people's minds, however, it is quite a stretch from making broccoli out of wild cabbage to making humans out of amebas, or, for that matter, out of mud, although the first man Adam is described in the Bible as being made of clay. Evolutionary theory does not claim that complex creatures and plants are assembled anew by random encounter between molecules from the environment. Instead, evolutionary theory, including that addressing problems of life's origin, always invokes both a pre-existing condition and a mechanism that can be demonstrated to operate today in the lab or in nature. The pre-existing condition can be anything known to exist on Earth, or anything that existed in the past provided there is tangible evidence for the condition. A single large continent, for example, can be considered a past pre-existing condition of Earth, and there is overwhelming evidence from the field of geology that such a single continent, named Pangaea, existed 250,000,000 years ago.

Pre-existing conditions also can refer to chemical makeup of mud, water, and atmosphere, to a living community of plants, animals, fungi, and microbes. A fauna consisting of amphibians and reptiles, for example, is a pre-existing condition. Scientists who study life's origin tend to start their research with mixtures of chemicals known to exist throughout the universe, including on Earth in its infancy. Scientists who study the diversity of modern organisms tend to start their research with questions that can be answered using comparisons, typically with fossil or presumably related species.

These scientists might focus on some group like minnows or grasses, and ask: in what ways are these species similar in spite of their differences and in what ways are they different in spite of their similarities? The underlying assumption is that similarities exist because of common origin and the differences exist because of evolutionary divergence.

The differences and similarities are used to group species according to relatedness, but such groupings actually are hypotheses to be tested by further research using new technology when and if such technology becomes available.

Both groups of scientists mentioned above are studying evolution and neither is assuming a supernatural force at work. Those scientists trying to discover how life originated will ask whether certain combinations of non-living materials can assemble spontaneously and then exhibit life-like functions. An example of such observations could be spontaneous formation of vesicles that end up internally concentrating certain chemical compounds and excluding other chemicals. Such vesicles may not be living but they do exhibit a fundamental property of life, namely, the ability to sequester and exclude compounds, that is, to regulate their internal environments. In developing such vesicles, scientists demonstrate that physical entities exhibiting fundamental life properties can arise without action by pre-existing life forms. Such research establishes plausibility but does not prove that life arose from non-living sources.

Proof of life's origin—exactly when and how it came about—probably is impossible, but demonstration of plausibility is highly possible and being accomplished even as you read these pages. The experiments are sophisticated, they involve complex chemistry of organic molecules, and in order to truly understand them, one needs a good background in physics as well as chemistry. Few citizens, and even fewer elected officials, have adequate knowledge of chemistry and physics to appreciate the findings of those scientists working on life's origins.

From a position of ignorance, it is quite easy to simply dismiss a realm of intellectual endeavor, especially if the products of that endeavor contradict what you've been told since early childhood. Nevertheless, chemists and physicists have done enough exploration of molecular-level origin-type events that can occur in abiotic (non-living) environments so that they are not very skeptical about the possibility of life originating without a supernatural hand. In fact, their work suggests life may be fairly common throughout the universe. This research can be found in a number of highly respected scientific journals, including *Proceedings of the National Academy of the United States of America, Origins of Life and Evolution of Biospheres, Astrobiology, Journal of Theoretical Biology*, and others.

By citing these journals, I'm not demanding that Joe Sixpac and Sarah Hockeymom read and understand them, nor am I intending to demean either those particular people or the groups they symbolize. I am, however, pointing out that the world's scientific establishment—that group of nerds who brought us the Green Revolution in agriculture, space stations and inter-planetary exploration, modern medicine, computers, software, petroleum, synthetic fabrics, gene therapy, DNA-based forensics, laser-guided missiles, and nuclear weapons—is the same group of nerds who tell us that Pangaea started to break up during the late Mesozoic, that humans and chimpanzees are at least 98% genetically identical, and that perfectly good complex carbon-based molecules, the same ones found in our own body, form in non-living soup from materials common throughout the universe.

No matter what the realm of human creative endeavor, be it art, music, literature, or science, the products of that endeavor usually, if not always, have impacts beyond their immediate existence. That is, someone always asks what the products can tell us about ourselves. In the case of evolutionary biology, scientists are telling us plenty about ourselves. Some of us are intrigued by the story of our past, others are disgusted. The intrigued ones sometimes become scientists themselves; the disgusted ones sometimes become politicians.

The world's scientific establishment is exceedingly intelligent, collectively speaking, and constrained by a rigid and ultimately unforgiving evaluation system known as "peer review" in which research is scrutinized in detail for flaws in design and execution, interpretation is constantly questioned, and studies that seem to yield important discoveries tend to be quickly repeated. The repetition uses published methods and often is done by scientists competing with, or attempting to discredit, the ones who originally publish such discoveries. Any work, especially important work, that cannot be repeated by described methods is immediately suspect and the scientists involved can suffer loss of reputation if not outright estrangement from the scientific community and loss of jobs. Peer review is not infallible and mistakes are made as well as overlooked by reviewers. But in the long run, if the scientific work is important to others, then those mistakes will be uncovered.

Evolutionary biologists are no different from cancer researchers or people studying ways to make biofuels out of yard clippings in terms of their use of, and being subjected to, the peer review system. The word "evolution" as seen and used throughout both popular and scientific literature therefore infers not only the fact of its occurrence, the processes involved, and the underlying theory, which is the central unifying theme of biological science, but also the massive research endeavor focused on that fact and those processes. A century ago research focused on structure and development—"morphology" and "embryology" in biological terms—and the ideas that guided that research stemmed largely from comparisons between supposedly related species. Today, however, although the comparative method remains central to this research, biologists go straight for the genes.

As indicated above, evolutionary events and relationships as diagrammed in textbooks, and in the underlying scientific literature, actually are hypotheses to be tested. In this regard, evolutionary biology is no different from any other biology, or any other science, for that matter. In science, all discoveries, ideas, claims, and paradigms are up for verification and, in many cases, modification. The way science proceeds to test such hypotheses varies, of course, between disciplines. In the case of evolutionary biology, hypotheses tend to be tested by using criteria and characters to establish relationships and recover events, but these are characters and criteria that were not used in development of the original hypothesis.

For example, relationships proposed based on structure are often tested using genetic methods not available when these relationships were first proposed. Evolutionary biologists have been almost as opportunistic as applied biologists in adopting new technologies as tools to investigate age-old questions. And what new technologies tend to provide are new ways of making observations, or of observing features not visible with previous technologies. Developing technology, regardless of its form, tends to provide the scientific community, including evolutionary biologists, with the needed "characters and criteria that were not used in development of the original hypothesis."

In the late 19th Century, for example, a biologist might easily conclude, based on comparative anatomy, that humans were far more closely related to chimpanzees and gorillas than they were to tailed monkeys such as green guenons and macaques, or, for that matter, to other vertebrates such as mice and chickens. This 19th Century biologist might express such a hypothesis as an evolutionary tree showing humans and great apes occupying some of the top branches, usually in the upper right, and other vertebrates occupying lower branches, with the branches themselves reflecting lines of descent. Indeed this kind of evolutionary tree is exactly what was presented to the public by Ernst Haeckel, the German scientist who embraced and exploited Darwin's ideas, spending much of his illustrious career exploring the implications of natural selection (see Gould's *Wonderful Life* and the Wikipedia site for a short discussion of Haeckel and examples of his drawings). Often, however, these Haeckel-type trees were seen as "facts" rather than as hypotheses, and the accompanying subtle message of progress (humans are at the top of the tree) was inescapable.

In any discussion of evolution, and particularly its scientific underpinnings, it's vitally important to remember that genetics as a discipline did not exist until after the re-discovery of Mendel's work in the early 1900s, nearly half century after publication of Darwin's *Origin*. Immunology, with its indirect techniques for assessing similarity of protein structure, did not originate in recognizable form until the 1890s and the full appreciation for species diversity, within the field of immunology, did not truly enter the scientific conversation until mid-20th Century. DNA was not established as the carrier of genetic information until the 1950s, and modern sequencing technology that suddenly makes all kinds of evolutionary hypotheses testable, became widely available only after 1990.

The mathematical algorithms used to construct phylogenies (evolutionary hypotheses) were developed primarily in the 1970s and widely accepted, with many modifications and broad applications beyond biology, for example, into linguistics, shortly thereafter. Over the past century, therefore, evolutionary biology has seen a massive growth in technology appropriate for the testing of hypotheses regarding relationships and descent. The eager use of that technology by scientists has only strengthened the fundamental theory, making it far more deeply embedded in, and intertwined with, biology, medicine, and agriculture than Darwin could never have imagined.

Nowadays, those scientists comparing minnows or grasses, trying to determine relatedness in order to recover evolutionary histories, routinely test their hypotheses using nucleotide sequences in DNA. The basic structure of DNA is rather commonly known, even by non-scientists, usually as a result of reading about the molecule in criminal cases or watching crime shows on television. The O. J. Simpson murder trial (1994-95) provided the American public with probably the most extensive lesson in molecular genetics ever shown on public television up to that time, and Simpson's defense attorney, Johnny Cochran, did a masterful job of dissecting the DNA evidence, especially focusing on the potential mistakes, flaws, and technological glitzes accompanying use of such genetic information.

Molecular technology is much more sophisticated and reliable today than it was in 1995, and law enforcement agencies are well aware of the various uses of DNA in criminal investigations. There's no better demonstration of the proverbial two-edged sword in science than the Third Millennium molecular technology: applications for "good" are abundant and widely accepted, with the notable exception being some localized fear of genetically modified crops. But if you're a creationist, applications for "evil" are just as abundant, and these applications strengthen evolutionary biology immeasurably. Evolution occurs. All organisms evolve. Even individual genes evolve. Evolution is a fact. People lie or perpetuate ignorance for political reasons, but molecules don't.

This discussion of DNA is relevant to the question of "What is evolution?" because mistakes in the making of DNA, by organisms, produce the mutations that have been called the "raw material of evolution." What is DNA and what are mutations? DNA is a very large molecule (chemical substance) consisting of two strands of nucleotides joined together. Nucleotides are molecules, with names we abbreviate by their first letters: A (adenine), C (cytosine), T (thymine), and G (guanine). Thus a single strand of DNA might have the sequence ACTTGCATCGTAC. A different strand might have the sequence GCCTACGATCGT.

It's not the fact of these nucleotides being present, but the sequence in which they occur that makes the two strands different. In this way, with the "information" being contained in a *sequence* of characters, DNA is more or less analogous to the writing in this paragraph. Different sequences contain different [genetic] information just like this sentence contains information that is different from any sentence in the previous paragraph.

Mutations can take numerous forms, but four common ones are insertions, deletions, substitutions, and inversions. In the first of the above sequences—ACTTGCATCGTAC—an insertion could be ACTGTGCATCGTAC, a deletion could be ACTTCATCGTAC, a substitution could be ACTGGCATCGTAC, and an inversion would be ACTTACGTCGTAC. The effects of changes are illustrated easily by a little Biology 101 exercise involving the University of Nebraska's slogan—GO BIG RED. The evolutionary exercise is as follows: Make ten sequential changes (that is, carry out evolution). You can use any of the above mutation types, the only condition being that the result needs to make sense, that is, survive in the environment known as the human mind. Here is an example of a full credit answer:

GO BIG RED
NO BIG RED
NOW BIG RED
SNOW BIG RED
SHOW BIG RED
SHOW BIG REED
SHOW BO REED
SHOW NO REED
SHOW NO DEER
SHOW NO DEAR
SHOW NO FEAR

The main insider knowledge you have to have is that "BO" is Bo Pelini, the new football coach during the semester this example was used. There were a number of other successsful evolutions of this slogan, some of the end points being:

HOT PIG PEN
NO BAG DEER
GO BY ED
TOO BAD ROD
GET BUGS FED

ONE BIG CART
DOTS ON DEAR

It would be an interesting exercise to take all these end points, subject them to some phylogenetic analysis using commonly (and freely) available relationship-revealing software, then construct an evolutionary hypothesis. Substitute the letters A, C, T, and G for those letters used in the above examples, remembering that in DNA even punctuation has a code, and the fundamental events of evolution become clear: evolution is first and foremost a molecular phenomenon with manifestations in phenotype, that is, in the equipment any organism possesses for finding food, finding a mate, and utilizing, or protecting itself from, its environment.

Macroevolution has, ever since publication of *The Origin of Species by Means of Natural Selection,* been a major problem for creationists, and probably understandably so. How is it possible for the evolutionary processes we now understand to generate the large and inclusive, but highly variable, categories of organisms? In other words, how can whales, horses, bats, and dogs all evolve from small mammals over a period of tens of millions of years? How is it possible for hawks, hummingbirds, robins, and warblers to evolve from dinosaurs? How is it possible for wheat, corn, and soybeans to evolve from algae?

These questions are, admittedly, challenging ones, although the molecular biologists and geneticists have answered them fairly straightforwardly, especially over the past thirty years. The answer involves various categories of genes and the diverse mechanisms of expressing those genes, an area of biology that continues to amaze even those actively engaged in it. The answer also involves discovery of events and processes that make macroevolution highly plausible, removing evolutionary scenarios from the realm of speculation and placing them into the realm of testable hypotheses.

The main culprits in this story are the *homeobox genes.* These genes control development, that is, the acquisition of adult form, especially in multicellular organisms such as fruit flies, mice, and humans. The *homeobox* is a DNA sequence within these genes, and all the genes that contain this sequence are considered members of the *homeobox gene family*. A gene is said to be "expressed" when a cell uses the DNA sequence information in that gene to actually direct the structure of a polypeptide, that is, a chain of amino acids, which can then become a functional protein. Proteins come in many sizes and shapes, of course, depending on the genetic information that was used to build them, and thus the sequence of their amino acids. Consequently, proteins carry out an equal diversity of functions. Proteins made from homeobox genes, however, function as *transcription factors*. The phrase "transcription factor" refers to a regulatory protein; that is, transcription factors regulate the expression of other genes. In general terms, then, homeobox genes control development.

Body proportions are the result of developmental events, so mutations in homeobox genes can affect not only development, but as a result, body proportions. Small changes in body proportions produce structural diversity which the human brain tends to interpret as difference within the context of similarity or relatedness. Dogs are a very familiar group of closely related individuals that can and do differ in body proportions (compare a greyhound with a dachshund). Tigers and house cats also are structurally similar but differ in body proportions. All of the magpies, blue jays, crows, and ravens belong to a single family, yet when seen in a zoo or in the wild, obviously differ in body proportions. Similarly, so do all the ducks, geese, and swans. Most experienced gardeners can distinguish between oak species, based on leaf structure, and in doing so use proportions of leaf lobes and points to make such distinctions.

But large changes in body proportions result in macrodiversity, thus are macroevolutionary events. The discovery of regulatory genes resulted in an explanation for the origin of large changes in body proportions. The discovery that these genes are present in all multicellular animals makes this explanation plausible. So we can ask, and answer, the following questions:

Do genes exist that regulate body proportions? The answer is "yes."

Are these genes widespread and common to all animals? The answer is "yes."

Is macrodiversity mainly a matter of body proportions? The answer is "yes."

This list of evolutionary questions, to which the answer is "yes," could go on for at least a page or two. So it's important to remember that an organism's genes are largely responsible for its development, including embryological or larval life, and the environment in which development occurs can potentially select for or against mutant genes expressed during this time in the early lives of individuals. There is no real conflict between nature and nurture; they work together, often in complex and little understood ways, to shape an organism and provide it with the ability to survive or, alternatively, to deny an organism such ability. And evolution does not work only on adults. Indeed, larval forms have long been considered raw material for the production of biological innovations by evolutionary forces.

For example, in the early 1900s, Walter Garstang, an invertebrate zoologist at Lincoln College, Oxford University, proposed that modern chordates evolved from sea squirt larval forms and supported his hypothesis with extensive observations on early development of a wide variety of species. Unlike many if not most of his contemporaries, and his intellectual descendants, he expressed his ideas in poetry as well as in the scientific literature. Garstang's basic idea, namely that larval (immature) stages of species are prime sources of evolutionary innovation, has kept several generations of biologists busy. Nowadays, of course, those biologists are using molecular methods to test their hypotheses.

The take-home message from all the foregoing discussion is that biologists now know a lot about how genes control development, as well as how developing organisms interact with their environment. This whole realm of scientific endeavor is sometimes referred to, by scientists themselves, as "eco-evo-devo" although not necessarily in that order. Darwin would be amazed. Suddenly, in historical terms, *The Preservation of Favoured Races in the Struggle for Life* seems like a rather simplistic explanation for the enormous diversity of organisms—from bacteria to primates—that inhabit this planet. The simplicity does not negate its validity, however; virtually everything science has accomplished since November 24, 1859, the *Origin of Species'* publication date has strengthened, supported, and expanded upon Darwin's basic claims, as follows:

Are more individuals produced within a species than can survive? Yes.

Do individuals within a species vary genetically? Yes.

Do some genetic variants produce relatively more offspring than others? Yes.

Those three questions, and their answers, are where discussions of evolution must begin. Such discussions are nowhere near ending, but they have gotten progressively more sophisticated, complex, and convincing, almost on a daily basis. Evolution is simply a fact of life, a fundamental property of life, and probably on every planet in the universe upon which life exists, regardless of its form.

11. What Kinds of Organisms Share the Planet with Us?

> *When it became clear to me that a book of this sort would help to acquaint Oklahomans with the wealth of their state's birdlife, I was faced immediately with the problem of which fifty birds to include.*
> —George M. Sutton (*Fifty Common Birds of Oklahoma and the Southern Great Plains*)

What kinds of organisms, besides humans, live on Planet Earth? This question may not seem like a particularly relevant one for a book entitled "Intelligent Designer," but it actually is the central question of such a book. We might rephrase the question, for example, as: What did God really build when He made the Earth in seven days? Once we begin to assemble an inventory of Creation, then the answer to this particular question gives us a sense of God's power because that Creation turns out to be far more diverse, complex, and wondrous than we imagined. We might also ask an equally related question: Where did all this diversity come from? An inventory of organisms sharing our planet not only turns out to be extraordinarily large, it also by its very nature tends to suggest the processes that produced it.

All true scientists and most well-educated non-scientists today accept the fact that evolutionary processes worked to produce this diversity. Such acceptance is consistent with the famous geneticist Theodosius Dobzhansky's assertion that "nothing in biology makes sense except in light of evolution" (see essay with this title in *American Biology Teacher*, 1973, 35:125-129). Nothing that has happened in biology since Dobzhansky make that claim has undermined it. Indeed, research in molecular biology, which came to age after 1973, has greatly strengthened his assertion.

Practicing scientists who try to learn about the diversity of even a few groups of organisms soon realize that unless they choose some group known to be pretty impoverished, species-wise, they're in for a lifetime's work that will still leave much to be explored. And those who try making sense of a really diverse group, for example, beetles, soon end up specializing on a relatively small subgroup, maybe a family with only 5,000–10,000 species. But in both cases, these modern scientists are replaying Darwin's experience on the *Beagle*. Exposure to real biological diversity, not just the few plants in your yard, or the large pets sleeping by your fireplace, inevitably leads one to an evolutionary explanation.

For elected officials and other culture warriors who are not biologists, or haven't studied organismic biology, here are some simple rules and terms for understanding life's diversity. Don't be too put off by a few words that sound like something out of a textbook. If you understand terms like *bailout, hedge fund, 501(c)3, 401k*, and *Afghanistan*, then the paragraphs below ought to be a piece of cake:

(1) Scientists group known living organisms into several categories that share traits. Such shared traits may be structural (for example, the similarities between dogs, coyotes, wolves, foxes, etc.), or molecular (for example, as in bacterial DNA), or both.

(2) These groupings are arranged in increasingly inclusive categories, again based on shared similarities. For example, dogs, wolves, and foxes can be put in a larger group of carnivores ("meat eaters") that includes weasels, skunks, and bears, based on tooth structure and other features.

(3) The science of grouping, or classification, is called taxonomy, and the groups themselves are called taxa (pl; singular = taxon). A scientist who is studying the way various species of a group are related, for example, is called a taxonomist. It's impossible to even begin understanding the

diversity of life unless you also know about, and understand the rationale for, grouping organisms.

(4) The most inclusive groups [of organisms] are called domains. There are three such domains: Bacteria (or Eubacteria), Archaea, and Eukaryota. The names of these domains, like names of all taxa, are proper nouns.

(5) Each domain has one or more kingdoms. Domain Eukaryota, for example, includes kingdoms: Protista, Fungi, Plantae, and Animalia.

(6) Each kingdom also has some subgroups. Kingdom Animalia, for example, includes groups known as phyla (pl; singular = phylum) (Mollusca, Arthropoda, Chordata are good examples). Animals within a given phylum share a basic body plan; that is, they all have the same fundamental architecture.

(7) Phyla have subgroups known as classes, classes have subgroups known as orders, orders have subgroups called families, families are made up of genera (pl.; singular = genus), and each genus includes one or more species. Most freshman textbooks illustrate such a system with a complete classification of some familiar organism such as a house pet. For example, your cat is:

Kingdom: Animalia

Phylum: Chordata

Class: Mammalia

Order: Carnivora

Family: Felidae

Genus: *Felis*

Species: *Felis catus* Linnaeus, 1758

(8) Note that in this example, the genus name (*Felis*) and species name (*Felis catus*) are italicized, and all other taxon names (family, order, class, phylum) are proper nouns (capitalized) but not italicized. Linnaeus, the great Swedish taxonomist who devised our modern system of naming plants and animals, called the housecat *Felis catus* in volume 1, page 42, 10th edition of his famous book *Systema Naturae*,

published in 1758. This short discussion may seem like a rather arcane bit of old science, but in truth it's the first lesson in reading about living organisms, namely, a lesson in what the words actually mean and imply. If you get into scientific literature, you'll find that there are numerous sub-taxa (subphylum, subclass, etc.) in addition to the ones given above.

(9) Instead of structural features like retractile claws (for example your cat, lions, and tigers), or non-retractile claws (your dog, wolves, coyotes, and foxes), Bacteria and Archaea are classified mostly by biochemical and genetic traits, as well as their ability to grow on various media (usually "soups" made of proteins, sugars, vitamins, and salts).

(10) Bacteria are very tiny, occur in staggering numbers, and equally staggering diversity, everywhere on Earth. Archaea look superficially like Bacteria but they are different biochemically and tend to live in inhospitable environments like hot springs. There are several thousand known species of Bacteria and Archaea and observation of bacterial diversity in places like the ocean, based on modern molecular technology, suggests there may be several million species yet to be discovered. Also, biologists are having a lively discussion over what actually constitutes a bacterial species.

(11) Protista (familiar amebas, paramecium, algae, etc.) are extraordinarily diverse and complex, too, although you can't really see that complexity without using the electron microscope, an instrument that can magnify specimens thousands of times. There are at least 100,000 species of known Protista, and probably at least a million undiscovered ones. As in all groups of organisms, the estimate of undiscovered Protista is based on the ease with which you can discover a new one if you make the effort. With Protista, it's pretty easy to discover a new species if you really try and have some idea where to look.

(12) Fungi are exceedingly common; there's a pretty good chance you're breathing in fungal spores even as you're reading this paragraph. There are approximately

70,000 known species of Fungi, with about a thousand new species being described annually. Mushrooms are, of course, Fungi, as is yeast used in baking. In addition to their economic benefits as food, Fungi also play a major role in the breakdown of biological materials in nature, thus returning nutrients to the soil.

(13) There are about 300,000 known species of plants (Plantae), with new ones being discovered every year. About 20,000 of these species are orchids, and another 10,000 species are grasses such as corn, wheat, and bamboo. Most plants are Angiosperms, or flowering plants, and their classification is based largely on flower structure.

(14) Scientists know about a million known species of animals (Animalia), and are discovering new ones regularly. More than a quarter of these species are beetles, although there are about 100,000 species of bees, ants, and wasps, another 100,000 species of clams, snails, and squids, and nearly 9,000 species of birds. The average size of an animal species is about half an inch, which means that human beings, if they notice other species at all, are looking at very large ones. A typical house cat is about 10 pounds, or about 4.5 kilograms. If the average animal is a beetle half an inch long, then the cat weighs at least 5,000 times as much as the average animal, and probably has at least 5,000 times the volume. In general, the American public is notoriously ignorant about animals, especially the average animal half-inch or less in length.

There is some question, even among scientists, over whether viruses really are alive and should be counted when considering life's diversity. I take the position that yes, viruses should be included in the planet's inventory, if for no other reason than their economic importance and their intriguing interactions with other organisms. Viruses cannot survive apart from Bacteria, Archaea, and Eukaryota, because viruses must infect other organisms in order to complete their life cycles. Collectively, the world's viruses contain an enormous amount of genetic information in their

DNA, information that is moving all around on the planet, sometimes getting incorporated into other organisms' DNA, and generally surviving quite well regardless of whether scientists consider it alive or not. As you might therefore suspect, viruses also evolve, many of them quite rapidly.

Most species live in the tropics, that is, between 23° 26′ 22″ north and 23° 26′ 22″ south of the Equator, that is, between the Tropic of Cancer and the Tropic of Capricorn. Scientists estimate that about 70% of all genetic information that spells life on Earth is found in organisms that live in this area. Tropical forests are especially rich in all kinds of species, from microscopic bacteria to giant animals like people, chimpanzees, and tigers. Humans are currently destroying tropical forests at the rate of approximately 50 acres a minute. We have been doing this destruction now for several decades and are likely to continue doing it until most of the genetic information that spells "Life on Earth" is gone, that is, extinct.

All of us, including the American President, all United States Senators and Representatives, all 50 state Governors, and every elementary school child in America, all are living in, and contributing to, an Age of Mass Extinction that ultimately will dwarf the Permian Extinction, which occurred about 250 million years ago and resulted in the loss of about 90% of all genera known from the fossil record. It remains to be seen whether human life will continue as we currently experience it in the United States, when most of Earth's genetic information is gone—forever; extinct—along with petroleum, and the human population is stabilized at about nine billion (as predicted by scientists).

I once heard a field man from the Nebraska Game and Parks Department refer to the four main categories of fish as "game fish, trash fish, chubs, and shiners." I thought at the time this classification scheme was highly practical, regardless of how un-biological it was. The practicality was based on use: food (game fish), fertilizer (trash fish), and bait (chubs and shiners). For those who prefer such a classification to real ones (domains, kingdoms, etc.), I propose the following to include not just fish, but all organisms: germs, food, pets, and vermin. Germs make us sick; food keeps us alive; pets keep us happy; and vermin give us something to kill, ignore, or talk about.

Vermin are the biggest problem for elected officials and other people in positions of power who take pride in remaining ignorant of life on Earth. The reason vermin are a problem is this: the vast, indeed the overwhelming, majority of what most people call vermin ("bug," "worms," etc.) are tiny, inconsequential in human terms, and not at all dangerous. Even most spiders, which tend to terrify the uneducated, are pretty harmless to humans. Vermin, however, are exceedingly interesting, lead fascinating complex lives, and usually deal successfully with the wild environments Earth has to offer.

For these last reasons, vermin usually interest kids, or people who never outgrew their curiosity about nature. Most humans, especially unhappy ones in urban areas who tend to vote for candidates who promise relief from whatever misery afflicts us, outgrow their childlike curiosity about the same time they reach puberty. This loss of curiosity means that ignorant elected officials are most pleasing to ignorant, and incurious, constituents. Ignorance and lack of curiosity are not healthy traits for a highly complex, energy gulping, resource consuming, military worshipping, and technologically dependent society such as ours.

What's the easiest and quickest way for a responsible person in a position of social power to combat such ignorance about the world's biota? The answer is simple: go to the natural history museum and be seen there, looking fascinated with whatever is on display, especially if it's nonhuman. You don't have to actually *be* fascinated; you just have to *look* fascinated. (Most elected officials, especially if in their second terms, know full well the difference between *being* and *appearance*.) Shells and fossils are terrific for this purpose. You can just hear those voters saying: Mayor

_____ loves shells and fossils! Wow, I thought they were cool, too, when I was a kid! Then in the back of this person's mind is the nagging thought: maybe I should still be curious about shells and about the natural world in general. After all, those kids who love shells and fossils grow up to be voters.

Periodically you see the term "ecosystem services" in newspapers and magazines. What does "ecosystem services" mean? The "services" part should be easy; plumbers, electricians, and auto repair shops provide services; that is, they do things for you that you cannot, or will not, do for yourself. "Ecosystem" may sound a little more scientific, but the word actually refers to something familiar to many of us, namely, an identifiable landscape that is relatively stable and regulated by natural processes.

A large area of coniferous forest (mainly pine trees and their relatives) can be considered an ecosystem. The natural processes that regulate a coniferous forest are birth, growth, death, decay, decomposition, movement of water and soil nutrients, predation (among animals that live in the forest), and the drifting of continents. Similarly, those same kinds of processes regulate the life of coastal marshes, prairies, rivers, and deserts, but the species that live in these various ecosystems vary.

Ecosystem services are ones provided by natural environments, for the common good, and usually are services that either individual humans, or small groups of humans, cannot provide. In this sense ecosystems are like armies and police forces. Hurricane Katrina in August, 2005, probably demonstrated the value of ecosystem services better than any other recent event, mainly because humans, for decades, had ignored the need for coastal wetland services, thus setting up conditions that eroded the Mississippi Delta, in the process destroying the ecosystem whose services they so sorely needed.

Those services included support of a commercial and sport fishery, protection from flooding produced by hurricanes, and feeding of wildlife. The wildlife in turn supported a tourist industry. Most of these services were provided by mud, vegetation, and silt from the Mississippi River. Before the introduction of atrazine into the American corn belt, construction of the New Orleans shipping channel, and other delta developments, nature, that is, the coastal ecosystem, provided the services. Humans declined the help of their coastal ecosystem, but Hurricane Katrina, followed closely by Rita, demonstrated the value of extensive coastal wetlands. Think needing an electrician after you've purposefully cut down the wires and poles, believing they were useless.

The word "ecosystems" implies the existence of many different organisms, from bacteria to large plant-eating and carnivorous animals such as bison and wolves, all living in a particular place on Earth and interacting in ways that over the long run sustain the system. Sunlight energy is captured by the plants, subsequently used by plant eaters of all sizes and kinds, whose bodies are eventually decomposed by organisms, for example, bacteria and fungi. Chemical elements such as carbon, nitrogen, and phosphorus are recycled through healthy ecosystems, and compounds like water have a global circulation pattern that affects local conditions.

Water that falls as rain in Pittsburgh could easily have been, at one time in its existence, flowing down a river in India. Similarly, carbon molecules found in the President's eye could easily have once been in a dinosaur, literally, as well as, perhaps, metaphorically. Nitrogen atoms, which make up about 80% of the air you breathe and are found in proteins of every organism, are captured by bacteria living in plant roots, converted to a form that can be used to make other molecules, and can ultimately end up in places ranging from a terrorist's trigger finger to a child's toe. None of this natural flow of sun energy and cycle of materials happens, however, without multitudes of organisms.

That is the take-home lesson: without those who share the planet with us—those millions of species of bacteria, fungi, algae and their relatives, plants, and animals—then Earth does not operate in a way that can sustain human life.

The underlying fact, however, and an irrefutable one, is that all these organisms have evolved from ancestors, bacteria and humans alike, and will continue to evolve until the Earth is burned by an exploding Sun about ten billion years from now.

12. What is Taught in Biology Class

> *We have noticed that in the germination of seeds the shoot grows straight upward into the sunlight and air just as uniformly and persistently as the roots grow downward into the soil.*
>
> —William Chase Stevens
> (*Introduction to Botany*, 1910)

There are a lot of answers to this question—What is taught in biology class?—depending on where and by whom the biology class is being taught. For the purposes of our *Intelligent Designer* discussion, however, we should probably assume that this class is General Biology for so-called "non-majors"—that is Bio 101—being taught at a college or university by someone well educated in the discipline. I pick the "non-majors" course because among the thousands of people who sit in these classes are to be found the future businessmen and –women, journalists, elected officials, and elementary school teachers.

Bio 101 is where the public meets biological science, and Bio 101 is very likely not only the first, but also the last, college-level science class this sampling of the public will ever take, at least if they can help it. Yet this bunch of citizens will be making big decisions for their nation, and a few of them will end up in positions of real power. So in this chapter I examine what our future leaders get exposed to, and in the next chapter examine what they should get exposed to, but in all honesty it's important to remember that exposure to biological science, regardless of its form and content, is not always a life-changing experience for 18-year-old future United States Republican Senators.

Although I've picked Bio 101 as the example for this chapter, remember that it is possible to get through the non-majors science requirements of most colleges and universities without being exposed to biology. There are plenty of places where Geology 101 works just as well ("Rocks for Jocks"), and although students don't often expect to encounter evolution in a geology class, they quickly discover that the paleontologists, some of whom teach these classes, really are mainstream evolutionists. And even if Geology 101 is not taught by a hard core evolutionist, it's pretty difficult to teach a geology class without referring to all those planetary events so central to the evolutionary story, for example, climate change, continental drift, volcanism, etc. But it is rare to pick up an introductory biology textbook without finding a section on human evolution complete with a phylogeny diagram (evolutionary tree) that includes both humans and apes. Rocks for Jocks texts do not contain such pictures. So biology class is the place where science and religion are most likely to meet head on, especially if both subjects are defined so narrowly as to mean human evolution and human creation.

A typical list of topics for General Biology includes biological molecules (starch, proteins, fats, DNA), cell structure and function, genetics and inheritance, a little bit of evolution, and often some "ecology" which is highly paradigmatic and stereotypical (carbon, nitrogen, and water cycles, geographic distribution of major plant communities, and prey-predator relationships). Depending on the institution, such a course may include a great deal of human biology, especially heart function and reproduction, because young people like that kind of material and it's "tough" but "useful."

By "tough" I mean the vocabulary can be challenging, as can all science vocabulary, and by "useful" I mean the material feeds the narcissism that so characterizes our human species. Everyone knows someone with heart trouble and people in general love to hear about sex and babies. Also, by talking about humans, an instructor can avoid a whole lot of other, relatively unpleasant but highly important subjects like biodiversity and conservation biology, in which the vocabulary is even more challenging than in physiology. Finally, by focusing on what students love, an instructor's teaching evaluations often can be improved over what they would be otherwise. Such evaluations are always a source of either pride and ego support, if highly positive, or if negative, then depression, stress, and self doubt.

Biodiversity (= biological diversity, or consideration of the many kinds of plants, animals, fungi, and microbes on Earth) is an extraordinarily difficult subject to understand and especially to teach, so it rarely gets included in beginning biology courses although textbooks usually contain a section devoted to those organisms with which we share the planet. Biodiversity is a difficult subject because the vocabulary is so extensive and the words we use to describe the lives of fungi do not necessarily work when we try to describe the lives of fish or pine trees. But the daily news is filled with biodiversity information, ranging from squabbles over endangered species to reports of drug resistant infections and bacterial contamination in recalled grocery store items. During the time this chapter was being written, for example, my own local newspaper devoted almost half a page to *Dracunculus medinensis*, the guinea worm, mainly because it's the causative organism of dracunculiasis, only the second among well over a thousand known human infectious diseases that might actually be eliminated from the Earth during our lifetimes (smallpox being the other, although suspicions persist that either the Russians and/or the US have stockpiles of smallpox virus as potential biological weapons).

Dracunculiasis is only a passing curiosity for the vast majority of American citizens. A drug-resistant *Staphylococcus aureus* infection, however, also concerns biological diversity and if we know someone, especially our own child, who develops such an illness, then suddenly our interest in biology, and microevolution, too, increases dramatically. So depending on the institution or instructor, content of a beginning biology course may be skewed toward such "useful" information which, although students may not all realize it at the time, is only a vehicle for presentation of basic biochemistry, cell structure and function, and genetics. For example, my own lectures for the first couple of weeks often address HIV and AIDS for three main reasons:

(1) By covering certain medical stories, the media routinely does public education on these subjects by default, so my students usually have at least heard the words and have learned a little bit about the subject before-hand;

(2) HIV can be transmitted sexually, and "sex with strangers" is a subject guaranteed to wake up any freshman class at any college or university; and,

(3) Textbook diagrams of HIV are labeled with basic vocabulary words (protein, nucleic acid, RNA, etc.).

Reason (3) is the most important of these three, although reasons (1) and (2) function to ease the task and provide drama or generate interest as well as making me appear to be concerned for students' personal futures. In addition, HIV usually is covered in a chapter on viruses, so that within a page or two of a labeled HIV particle can be found diagrams of life cycles and other viruses, including influenza, which then allow me to talk comparatively about biological materials, for example, different proteins and nucleic acids (= genetic information).

This comparative conversation is outright subversion and is aimed directly at students' so-called "habits of mind." In other words, I try to get beyond the definition of *protein* to a consideration of protein differences and the link between protein structure and genes. Thus a definition (structure of a protein) becomes only a means to understand role (function) and context (environment), as well as interactions between structure, function, and environment, that is, the essence of biology.

If my own experience is reflective of the national situation, then most beginning college students in a biology class want to know what they are "responsible for," which actually means "which facts I need to be able to recite." But biology is much more than a list of facts. Like all the sciences, it is a way of knowing, thus involves, or at least should involve, exercises designed to teach a person how to think about the natural world. Routinely, however, for practical reasons, Bio 101 focuses on the list of facts.

In today's higher education environment, and especially at large institutions, faculty members have neither the time, nor patience, nor resources to teach habits of mind to hundreds of freshmen sitting in a large auditorium. Nor, typically, do they have the imagination or the will to meet this challenge. Instead, what they have are publisher sales reps providing them with question banks, testing software, canned PowerPoint shows, and CDs or online tutorials. So biology as a discipline becomes highly descriptive, and almost pathologically paradigmatic; the epigraph on the opening page of this chapter is a typical example, although one chosen obviously for literary reasons. Nevertheless, I would not be at all surprised to see on some introductory biology exam a multiple choice question about the direction roots and shoots grow.

When the institution is a large one that is recruiting potential students like crazy (they all do, for financial reasons) and consequently faces the large numbers of successfully recruited incoming freshmen, then student and administrative desires clash in the arena of infrastructure. Most of the nation's colleges and universities saw this problem arising years in advance and if adequate financial resources were available, either built state-of-the-art multi-media auditoriums or converted existing lecture halls into electronic play rooms for faculty. My own university is probably fortunate in that it was either not willing or not able to build several monstrous lecture facilities, so it ended up with numerous smaller ones, in the 150-300 seat range.

A class of 250 freshmen is fairly easy to manage, especially compared to one of a thousand, so that it is possible, with a little bit of faculty idealism and ingenuity, to actually teach some biology to a significant fraction of the students. Thus when a faculty member is assigned to one of these classes, he or she has the option of being creative, which means investing the time and talent needed to change this three-times-a-week dog and pony show into a valid intellectual experience, or of using prepared textbook publisher resources, most of which are now online. In the vast majority of cases, they choose the latter.

Furthermore, student evaluations of teaching are an integral part of faculty performance reviews at many if not most institutions, and Peer Review of Teaching committees also are commonplace. At my own institution, faculty members are not supposed to be in the room when evaluation bubble sheets are completed, the forms are sent for analysis and summary without the faculty member touching or even seeing them, and when they get returned to the department, secretaries photocopy the written comments on the back. The data and photocopies end up in the faculty member's official file, and are used in promotion, tenure, and merit raise decisions. So the system is designed to produce homogeneity and promote use of highly standardized materials for which students pay, often dearly. Online supplemental study materials, sometimes with animations, CDs packaged with textbooks, PowerPoint shows consisting of textbook figures on a white or very light colored background, and study, quiz, and test questions tailored to each chapter, are now all fairly standard additions to the book itself.

So what gets taught in biology class is primarily content, especially such content that lends itself to right and wrong answers (think cell structure and function) as opposed to the meta-content lessons of professionalism, a respect for the scientific habits of mind, the behavior of scholars, and mentorship heavily laced with idealism (see the next chapter). America assumes that if one learns a certain body of knowledge well enough to answer multiple choice questions, or at least about 70% of them, then qualities such as respect for and dispassionate evaluation of evidence, and a habit of mind that asks for evidence to support an assertion about the public good, followed by critical evaluation of that evidence, will all ensue. Nothing could be further from the truth.

In fact, content, especially in a science class, tends to function in exactly the opposite way. A diagram of chemiosmosis, for example, just to mention one of the major topics of cell function, memorized along with the vocabulary necessary to answer a multiple-choice question about it, will not inspire anyone to question whether a coastal development policy will have long term consequences for tens of thousands of people. Instead, a diagram of chemiosmosis is very likely to have the same effect on a future elected official sitting back on the 29th row of Big School Auditorium that it has on me, the instructor standing at the podium in front of that same auditorium: surely biology has more to say to the world than this information, no matter how common and central the process might be to cellular life, and therefore to life in general.

Nowadays, I can look this term up if I have to; and I don't have to, today. I can acquire the vocabulary to understand this process at the introductory college biology course level if I need to; but this morning I don't need to. What I have to do, if I am this elected official, for example, is anticipate the environmental effects of development, hire some people who can and will tell me the truth about scientific and technical matters instead of whatever I want to hear, and convince the public that I am behaving in a responsible manner regardless of what they want to hear from me.

I, the teacher, however, am still faced with chemiosmosis as an integral part of chapters on cell biology in a $145 textbook for which 90% of these students will get $22 when they sell it back to the bookstore and who will also have the opportunity to answer the question—Was the textbook helpful?—as part of their formal but anonymous evaluation of me at the end of the semester. Furthermore, because this class is listed as "non-majors," it can easily end up enrolling people who are undeclared, or have opted for general studies for a year or two, but perhaps hold National Merit Scholarships or their financial equivalent and bring very high standardized test scores to the institution.

Routinely such people are perhaps peripherally interested in science or, especially, the health professions, but for whatever reasons have decided to avoid committing to a particular program of study. Maybe they just want to explore for a semester or two and have parents who are supportive of this decision. So the label on your course—non-majors—is code for "dumbed down, at least a little bit," but the clientele can be as mixed as the general population and include students who really ought to have skipped the introductory course and now be taking an upper division biology course instead.

Thus in an ideal world, what should get taught in biology class is a three-way compromise between content, context, and process. Much of the world of higher education is less than ideal, however, so a lot of what actually gets taught in biology class is whatever textbook publishers provide to instructors, including PowerPoint presentations, sometimes short video clips, or combinations of PowerPoint and audio tracks, and study questions at the ends of chapters. Is all this material "bad"? No. Is it "wrong"? No. Then what's the problem?

The problem is that this material, and the vehicles for its delivery, can very easily encourage the kind of disengagement that characterizes much of the student body in public higher education today. So a typical student might well say to himself or herself: I'll study this chapter, answer the questions, and see what happens on the exam, but right now I need to check in with my significant other and make some plans (or something to that effect). So whatever is being taught is mainly time management and compartmentalization of biological knowledge that in the real world outside the lecture hall is splattered into almost every aspect of life.

Whatever gets taught also is dictated in part by the numbers of students, the logistical burdens of testing, and a subtle but very real association between content and testing. This last factor is more important than many instructors and virtually all students realize. For example, it is very easy to ask multiple choice questions about cellular metabolism because there are lots of concrete processes to use and the answers are not very ambiguous and don't require much interpretation compared, say, to a phylogenetic tree based on molecular data and developed using mathematical techniques. Furthermore, if laboratory is involved, then experiments can be done repeatedly and designed to result in a "proof" or a tested hypothesis typical of normal science as envisioned by the general public. Thus from a pedagogical perspective, cell biology is relatively "clean," that is, uncluttered by ambiguity or vast groups of unstudied organisms. The same comment can be made about genetics.

Evolution, however, does not lend itself to such cleanliness; various interpretations are sometimes possible, and because evolution is an historical science, the nature of evidence differs quite a bit from that of cell biology. In addition, any presentation of evolution, especially at the freshman textbook level, leaves out many species, with "many" typically numbering in the hundreds of thousands. So in an ideal world, an exam concerning the central unifying theme of biology—evolution—should be one in which a student generates, in legible form, a document that reveals both knowledge and thought processes. In this case the knowledge should include not only evidence, but also techniques of study, and the thought processes should involve analysis, interpretation, and conceptual synthesis. And both the knowledge and the thought process must acknowledge the incomplete inventory we have of nature, on the one hand, but the enormous inferential power of the comparative approach, on the other.

Can a teacher carry off this feat, namely, the production of a scientifically literate individual who understands evolution and is neither frightened of it nor concerned that the world is going to hell if an elected official or a school board accept the concepts, observations, and implications of this unifying theme of biology? I believe the answer is "yes." But when the audience is several hundred freshmen taking a non-majors, general biology, "gen ed requirements" course in a large auditorium, the pedagogical techniques for accomplishing the task are not necessarily obvious, nor, perhaps, even yet developed.

So what tends to be taught in biology class are a few examples of molecular clocks and evolutionary trees (phylogenies) based on molecular evidence, especially that involving hemoglobin, and a brief description of the fossil record. Most often the species represented are vertebrates, for example mammals such as bears (the example in the text I am currently using). If an instructor has access to information on some of the truly glitzy examples like the mating songs of *Drosophila*, or HIV variation in African villages, and if the students are prepared for such material, then evolution can take on an immediacy that it would not have otherwise. But in the final analysis, a biology teacher is stuck with making sense of a staggering volume of diverse material.

Textbooks typically devote a number of pages, if not an entire chapter, to Darwin's voyage of the *Beagle* and his study of the Galapagos finches. Although this material is of major historical interest, it does little if anything to convey the idea of evolution, or the compelling evidence that molecular biologists have collected while testing phylogenetic hypotheses based on structure. In other words, one might ask: if humans and chimpanzees are so similar structurally and behaviorally, then should they also not be similar genetically?

This question actually is the basis for a molecular test of a hypothesis derived from morphology. The test fulfills the fundamental criteria of science, namely determination of whether a testable prediction is true. The null hypothesis in this case is one of no difference: humans and chimpanzees are genetically identical. But in rejecting that hypothesis, we discover that humans and chimpanzees share about 98% of their nucleotide sequences. The molecular test of a morphological hypothesis shows clearly that chimpanzees are our closest relatives; the evolutionary interpretation is that we share a common ancestor. What has happened, over recent decades, is that new technology has been used in an attempt to answer an old question, and this attempt ends up confirming the old answer.

The take-home message of all this previous discussion is as follows: in much, if not most, of American higher education, logistical constraints and publisher marketing strategies, rather than the fundamental nature of biological sciences, tend to dictate pedagogy. Thus what we teach is what large auditoriums with hundreds of 18-20 year olds and corporate media developers will allow and help us teach respectively. In the realm of biological science, is this educational experience appropriate for the audience that will, in a shockingly few number of years, become our teachers, doctors, attorneys, and elected officials? My answer is a "qualified no," "qualified" because I know from experience that some of these children will indeed acquire transferable skills necessary for a lifetime of responsible citizenship regardless of what an instructor throws at them, and "no" because the vast majority will not. Thus we must consider, as an alternative to the situation described above, what *should* be taught in biology class.

13. What Should be Taught in Biology Class?

> *Build* knowledge *of diverse peoples and cultures and of the natural and physical world through the study of mathematics, sciences and technologies, histories, humanities, arts, social sciences, and human diversity.*
> —Institutional Objective No. 2 (*Committee report from a large, public, land-grant university*)

"So what should we teach?"

This particular question has been asked by every school and every department that has ever offered a class, any class, including one in biology. The question also has been answered in about as many different ways as there are schools and departments, although for all their apparent variety, classes on a particular subject within a discipline turn out to be remarkably similar to one another. Schools in general, especially post-secondary ones, routinely assume they are serving their clientele and want to "teach" their students the information and skills necessary to take a subsequent step toward success in some kind of employment. Jobs in the health professions usually require knowledge of biological systems; so the answer to this chapter's question ends up being about the same as the answer to the previous chapter's question (*What is Taught in Biology Class?*). In other words, we rarely if ever ask two important corollary questions: (1) What should a person in a position of real power understand about how the living world really works? And (2) How easy is it for a pre-professional student to acquire knowledge and vocabulary necessary to take the next step toward a career in the health sciences?

The first of these questions, about appropriate knowledge for a person in power, arises from the fact that in a typical non-majors' science class one finds people who end up in elective office, or as executives in corporations, journalists, teachers, and spiritual leaders. In other words, that dingaling kid playing grab-ass, or trying to, with the girl next to him back in the 37th row of Olde Hall Auditorium while making no pretense of listening to your boring lecture on Mendelian genetics could easily end up your state's governor in 20 years. Or, he could be the son of an extremely successful entrepreneur and end up inheriting the family business, then deciding to spend $11,302,078 million of his personal fortune on a campaign for the United States Senate where, had he been successful he would have focused on banning gay marriage (do a Google® search on "Pete Ricketts"). So this kid needs more than Mendelian genetics; he needs to be as scientifically literate as he can be made to be in 15 weeks, given the circumstances surrounding biology class in Olde Hall Auditorium with 299 other people whose future cannot be foretold, especially by the professor at the front of the room.

The second question—How easy is it for a pre-professional student to acquire knowledge and vocabulary necessary to take the next step toward a career in the health sciences?—can be answered readily: very easy. If there is anything that is accessible to all humans in developed nations, it is knowledge. Every year I discard old textbooks that are at least 90% accurate, especially in the basics where they are probably 100% accurate, originally cost upwards of $150 each, and weigh about six pounds. If I could get them to a developing nation easily and cheaply, I'd do it in a flash. So it's an absolute lie that professors need to explain to smart kids what's in a textbook. Smart kids, who are the ones you want in the health professions treating you and your family, can learn this stuff on their own easily then ask questions about vocabulary and knowledge.

What those smart kids cannot get on their own, however, is laboratory experience. They need to get their hands on as many different kinds of organisms as they can, as early and as often as possible, and they need experience at original investigation. The exposure to diversity, especially of recalcitrant non-human species, and the struggle with research, leads to intellectual maturity and a certain respect for evidence instead of belief or desire. Furthermore, these experiences tend to erode arrogance, which is so often linked to belief and desire. So biology class should be focused on problems, preferably hard ones with multiple answers.

There is a third essential question, however, and it is discussed surprisingly often, although informally, by a few well-educated people, but only rarely does this last question get addressed formally by institutional action. This third question is a simple one: should a pre-professional student majoring in the health sciences be taught *biology*, as defined in the broadest sense, or should that student receive a highly technical and focused education?

Technical and focused educations are easiest and cheapest to deliver and are the least likely to generate controversy in the classroom. So this third question gets answered by default and college students, who are spending tens of thousands of dollars to get an education, and who all expect to get a job or acceptance to a professional school, tend to sincerely believe that technical and focused courses in the sciences are what they need to take the next step in their careers. Their parents also tend to believe this myth about the usefulness of an education. Unfortunately, what seems on the surface to be good for clientele seeking work in the health professions is not necessarily good for the country.

Perhaps as a way of illustrating this intellectual dilemma of deciding what to "teach" in biology class, I should introduce you to a typical first semester freshman biology class, Bio 101, at a large, public, mid-western university, namely mine. The list below is a summary of the types and sources of diversity among this sample of ~250 typical college students in a "non-majors" science course supposedly designed for "general education."

(1) Types and sources of diversity among typical mid-western state university students in a large (~250) introductory course.

Economic diversity: From very wealthy, well-dressed, and comfortable, to poorly dressed, struggling, and working at least one job.

Religious diversity: From absolute atheists to radical, charismatic, fundamentalist Christians, as well as Muslims, Buddhists, Catholics, Hindus, and a variety of sects within each.

Family history: From orphans and foreign adoptees to strong nuclear families with multiple generations interacting regularly.

Employment: None, to two, sometimes three, jobs, varying from library help to bar tender and waiter or waitress to hospital orderly, etc.

Military service: Combat veterans to peace activists, both men and women, including some in the National Guard or Reserves, and perhaps a few still on active duty.

Major, career goals: The full range; you name it, someone is majoring in it, or planning to make a career of it.

Co-curricular activities: None, a reclusive existence, to an overload of campus organizations and Greek houses, to scholarship athletes in revenue-generating sports.

Ethnic background: Far more diverse than most people realize. A typical large university biology class could easily have students from a dozen different countries, including those in Africa, the Middle East, and Latin America.

Gender and sexual orientation: Any large class (~250) will have flaunty heterosexuals, both male and female, as well as gays, lesbians, and probably some transexuals and bisexuals.

Appearance: Drop-dead, head-turning gorgeous to plain as a blank piece of typing paper to downright ugly to an appearance that obviously causes the individual some pain, especially given the show-biz standards to which young people are exposed nowadays.

Talents: As varied as those of the human species—superb athletes, artists, musicians, actors, and writers.

Drug and alcohol use: Absolutely none to hopelessly addicted, and everything in between.

Significant others: None, and proud of it, to obsessive and possessive to the point of being in constant cell phone contact, even (maybe especially!) during class.

Age and maturity: Seventeen to seventy, although the usual age it 18-40, with maturity not necessarily matched to age.

Reading skills: Superb to almost non-existent.

Superimposed on this diversity among students is an almost equally diverse, but more importantly pervasive and constant, exchange of information. This information comes in a wide variety of forms, in many media, and from almost every imaginable source—from immediate and sometimes intimate interactions with fellow humans to global news to images, sounds, and stories from imaginary worlds. The list below summarizes this array of information sources and content, which I call the "intellectual environment," using the term "intellectual" in its most general sense, namely, as anything having to do with the mind, thoughts, or actions stemming directly from thoughts.

(2) Information sources and content typically experienced by an American college student during any 24 hr period.

iPod, etc.: Personal music, voice audio, video, memos, of an incredible diversity and amount.

Podcasting: Commercial stations and media outlets providing free audio and video content, as well as digital audio files provided by instructors, institutions, etc., as part of a class or other educational activity.

Cell phones: Cell phone conversations are so common nowadays that they constitute a kind of background noise throughout much of our public space.

Text messages: These communications are now so common and frequent that the medium is being used as a public warning system.

Ring tones: An amazing diversity, used to customize one's cell phone, and used as a cash cow for telcom companies.

The Internet: This source of information and constant messages does not need to be explained. Indeed, words are inadequate to explain its impact on society, especially the younger people in developed nations.

Blackboard, etc.: Course management software, of which Blackboard is one example, multiplies the power of both students and teachers to communicate, design activities that can be done electronically, conduct online discussions, and do assessment.

Free downloads: There are hundreds, ranging from updates to existing software to new kinds of software that puts various kinds of powers in the hands of students.

E-mail: E-mail allows almost instant communication between any two people on Earth who have an Internet provider and access to a computer or smart phone.

Digital images: The number of these images that are made and transported daily must be astronomical, and the diversity in hardware used to make such images is almost indescribable.

YouTube, Facebook, etc.: Web sites that allow digital creativity and structured communication in a multitude of ways and are used heavily by younger people.

Electronic library access: Most colleges and universities subscribe to a number, often a very large number, of content providers that quickly make decades of original published research in all disciplines readily available.

- Sound bite reporting: "News" is highly structured, sometimes if not often in decidedly political ways, because there is so much of it and air time is exceedingly expensive. So in the media, an exceedingly complex world becomes highly abbreviated and simplified.
- Political stridency: Political discourse has become a theater of personal attack. It may always have been such, but information age technology establishes this type of discourse as "normal" because everyone does it.
- PowerPoint: This pervasive and ubiquitous software has taken away much of the incentive for in-class writing and creativity from both students and faculty; students watch the show instead of actively getting engaged with the material.
- Wireless networks: When a student can get online from anywhere on campus, then that student is inclined to get online and stay online with the illusion that he/she is actually becoming a scholar by doing so.
- Video games: This massive industry creates a virtual world that may have absolutely nothing at all with reality, at least as lived by a typical student.

So the mixture of heterogeneity (extreme, even in the heartland) and virtually indescribable complexity of the intellectual environment creates a challenge that American education has yet to meet successfully, at least on a broad scale, and especially in the sciences. We do an excellent job of training people for specific careers that require an ongoing, but nevertheless discreet, set of learning activities: physical therapist, physician's assistant, dentist, neuro-surgeon. We do a miserable job of educating people for life in a world that is rapidly becoming depauperate in natural resources such as water, petroleum, and arable land. And at the species level, on a global scale, we see the human mind starting to deteriorate: ethnic hatred, violence in the name of God, mass murder justified by tribal affiliations, and manufacture of truly horrible weapons of mass destruction by what is supposedly the most free and noble of nations (ours), all part of a war on terror delivered by angry young men with bombs strapped to their waists.

What *should* be taught in biology class, therefore, is respect and reverence for the natural world, a secular humanist view that we do indeed have power over our lives and futures, scientific literacy adequate to dispel one's fear of evolution, a clear sense of where humans fall in the grand parade of life on Earth, and a curiosity about natural phenomena that we cannot always see or understand.

Yet there are no textbooks written with these kinds of lessons, no multiple choice questions or PowerPoint presentations generated by the supplemental materials gurus at large publishing houses, and no graduate school courses designed to impart the idealism and skills necessary to teach such lessons. Instead, teachers teach what *should* be taught in biology class only by example, not by preaching or assigning paragraphs and chapters but by example, by tone of voice, choice of words, willingness to engage in conversation, willingness to help individual students deal with their mega-classrooms, and a determination to never, *never,* relinquish their humanity to a high tech information delivery console that resembles the bridge of a cruise liner.

Teachers who can transcend the logistical constraints of Third Millennium higher education in America usually are those who have themselves been taught in this fashion, who have been shown how to do it and asked to participate in the process, and who have gotten over the fear of not doing their job according to the expectations of some merit raise system.

One might consider the question: how much of the material in a typical freshman science textbook is actually retained by non-majors and ultimately put to use in making decisions that can have an impact on the common good? Will a detailed understanding of the citric acid cycle serve an accountant adequately when faced with local mayoral candidates who differ in their views regarding real estate development and zoning? Will the ability to accurately diagram transcription, post-transcriptional editing, translation, and assembly of multi-unit proteins serve a local businessperson adequately when considering his or her United States Senator's position on energy policy, given the fact that this Senator is up for re-election? The answer in both cases is "no," yet there is at least a 90%, maybe 99%, probability that citric acid cycle reactions, RNA synthesis (transcription), and protein synthesis (translation) were all given lots of pages in freshman textbooks and also covered on freshman, non-majors', biology class exams.

In all fairness to the system, however, we could also ask: do these exercises in energy metabolism and gene expression provide scientific literacy? The answer is a qualified "yes," qualified because scientific literacy arises not only from content, but also from use of content. If energy metabolism and gene expression are taught as examples of how to acquire biological knowledge and put that knowledge into a broader evolutionary and ecological perspective, then there is an excellent chance that at least some of the masses will come away from the student-professor encounter with enhanced scientific literacy. But it is incumbent on the prof to make some attempt at generalization, at leading the student from a "what do I need to know for the next multiple choice exam" condition to one in which the student is saying to himself or herself: this garbage is boring, and I'll never use it in my career, but my ability to acquire it, use it, and put it into context is a skill that could easily prove extraordinarily valuable to me in the long run.

The value of a humanist idealism is something transmitted vertically—from parent to child, teacher to student, and mentor to mentee. Mob behavior, however, is transmitted horizontally—student to student, businessman to businessman, and corrupt politician to corrupt politician. An individual teaching humanist idealism by example and actions knows that these lessons may not sink in for some time, even years, and that there are few reinforcing results to validate one's success. The mob, however, teaches by endorsement—

actions committed by the many, ostensibly on behalf of the many, are validated, and usually immediately and in large numbers.

The democratic ideal is thus easily subverted by charisma. If it's not deemed beneficial to society to have respect and reverence for the natural world, a secular humanist view that we do indeed have power over our lives and futures, scientific literacy adequate to dispel one's fear of evolution, a clear sense of where humans fall in the grand parade of life on Earth, and a curiosity about natural phenomena that we cannot always see or understand, then the mob will quickly erase any outward or official manifestation of these traits. And if such traits are deemed outright dangerous, as they seemingly were during much of the George W. Bush presidency, then a nation is in deep intellectual trouble.

Biology class must, however, have not only a meaningful approach and attitude, but also content, that is, facts. Regarding content, the best answer to this question—What should we teach?—is what might be called The Big Picture in Biology. The Big Picture addresses several major items, all of which have historical, scientific, and social significance, but the overarching theme is probably expressed best by Lawrence Slobodkin in his book *Simplicity and Complexity in Games of the Intellect:*

> "The natural world need not be logical in any obvious way. Science does not consist of imposing our reason on the world but rather reducing our preconceptions to the point that the world imposes its logic on us. This is very difficult indeed, involving a minimalization of our ego while maintaining our full powers of observation and receptivity. The capacity to perform this feat is what the teacher of science attempts to foster in the student. No one succeeds completely."

In other words, regardless of how much power we have, or eventually acquire over proximal events and conditions, the planet ultimately must be understood and our actions must ultimately be consistent with that understanding, else we become no different from the nearest beetle. The Big Picture items are:

(1) Earth is the only planet in the universe known to support life, so humans have an obligation to live within Earth's resources and do this living gracefully, with dignity befitting an animal in possession of such enormous intellectual ability.

(2) Life on Earth is characterized by enormous diversity imposed on great uniformity. The uniformity is to be found in the general architecture of DNA, common metabolic processes, etc.; diversity is in the massive number of species that occupy the planet.

(3) Evolution is the best general explanation *science* has for life's enormous diversity superimposed on great uniformity; that's why it's the central unifying theme of the discipline. And, there is simply a staggering amount of evidence in support of evolutionary theory as a general explanation for life existence and diversity.

(4) The vast majority of species that have ever lived are now extinct. It's real easy to be naïve and arrogant about our own, mainly because we're so smart, but the evidence to support the contention that most species are extinct is *very* convincing.

(5) Scientific evidence from the field of astronomy, geology, and biology strongly indicates that virtually all things in the universe have a beginning and an end, and our solar system is probably no exception. In fact, this finite life principle is uniformly accepted among scientists.

(6) The present distribution of life and other natural resources is a result of several billion years of planetary change (evolution, both geological and biologi-

cal). That distribution has significant social and political consequences, and so to some extent, our daily headlines are a result of planetary forces at work, forces over which we have no control, and producing events that we did not make happen.

(7) Science is different from technology. Both require fundamental knowledge, but technology seeks to *control* nature, whereas science seeks to *understand* nature. Control is not necessarily "good;" understanding is not necessarily "bad." It's what humans do with their control and understanding that make humans "good" or "bad."

(8) Many of our most difficult social and political problems have a major biological component: racism, sexism, unwanted pregnancy, global energy distribution, intellectual and cultural richness, the definition of "human being," narcotics, global water distribution, genetic "engineering" and its consequences, infectious disease evolution and transmission, our relationships with insects, etc. This list could go on for several more pages.

(9) We are surrounded by biological materials and information, and our lives are enriched immeasurably by being aware of the biology that infuses our daily lives, from gardens and landscaping to local backyard wildlife and pets, to biological patterns in clothing and jewelry.

(10) The scientific and technological explosion is not going away any time soon, so it's better to be educated and confident in your ability to understand scientific issues than it is to be scientifically illiterate.

The Big Picture as outlined above includes both commonly accepted facts ("The present distribution of life and other natural resources is a result of several billion years of planetary change." "We are surrounded by biological materials and information." "The vast majority of species that have ever lived are now extinct.") and interpretations of, or inferences derived from, those facts ("Many of our most difficult social and political problems have a major biological component." "Humans have an obligation to live within Earth's resources and do this living gracefully, with dignity befitting an animal in possession of such enormous intellectual ability.") Is such a Big Picture message ultimately of far more benefit to our society as a whole than widespread knowledge of gene expression? I contend that the answer to this last question is a resounding "yes."

14. What is the Meaning of "Scientific Literacy"?

> *When I was near the lowest point of my illness she sent me a wicked book by some evangelist—a word I have long used as a curse—about how Huxley will not look his (the evangelist's) substitutes for arguments in the face, how that geology supports the books of Genesis (which is a lie) how that the gospel of St. Mark was written before A.D.38 (which is idiotic) and all those dismal things.*
> —H. G. Wells (*Experiment in Autobiography*)

"Scientific literacy" refers to the ability to read scientific information, interpret graphs and figures, understand the evidence used to support arguments about scientific and technological issues, and evaluate sources of data used to support public actions. The so-called Weapons of Mass Destruction (WMD) information used by the United States government to justify the war in Iraq is a perfectly good illustration of evidence that could not stand up to scrutiny, even by the scientifically semi-literate. Evidence used to evaluate New Orleans levies and Mississippi Delta wetlands prior to Hurricane Katrina, however, was tangible, well documented, and subjected to considerable scrutiny. This evidence was real, credible, and right on target. In Katrina's case, it was the political machinery that failed its scientific literacy test over a period of decades.

When faced with a decision affecting the public welfare, a scientifically literate person does not ask: what do I *want* to have happen? or what do the *voters* want to have happen?

Instead, he or she asks: upon what basis should a decision be made and what is likely to be the aftermath of that decision, given what we actually know about the current situation or condition? In other words, scientific literacy imposes a certain level of rationality and intellectual honesty upon individuals, and hopefully, those in positions of real power.

An excellent description of the scientifically literate is provided by the National Academy of Sciences (National Science Education Standards. Copyright © 1996 by the National Academy of Sciences. National Academy Press, Washington, D.C.) To quote:

Scientific Literacy:

"Scientific literacy is the knowledge and understanding of scientific concepts and processes required for personal decision making, participation in civic and cultural affairs, and economic productivity. It also includes specific types of abilities. In the National Science Education Standards, the content standards define scientific literacy.

"Scientific literacy means that a person can ask, find, or determine answers to questions derived from curiosity about everyday experiences. It means that a person has the ability to describe, explain, and predict natural phenomena. Scientific literacy entails being able to read with understanding articles about science in the popular press and to engage in social conversation about the validity of the conclusions. Scientific literacy implies that a person can identify scientific issues underlying national and local decisions and express positions that are scientifically and technologically informed. A literate citizen should be able to evaluate the quality of scientific information on the basis of its source and the methods used to generate it. Scientific literacy also implies the capacity to pose and evaluate arguments based on evidence and to apply conclusions from such arguments appropriately."

Because this set of intellectual skills is derived from the National Academy's science education standards, it pertains most appropriately to students, so the individuals who are supposed to impart such literacy are teachers involved in science education. The target audience actually consists mostly of people from 5 to about 22, that is, those currently enrolled in school—kindergarten through 4-year colleges. All states have established standards and curricula for the basic subjects and if in compliance with current Federal legislation also have standardized tests for assessing student progress in various subjects, including science. *What our nation does not have is a scientific literacy test for elected officials.*

Nor do we have such tests for journalists, although programs in journalism and mass communication at our 4-year colleges do have "liberal education" requirements that include some science. Thus only after the fact do we ever learn whether elected officials have demonstrated scientific literacy when called upon to do so by circumstances, sometimes ones beyond their control, and such historical perspective is provided only if we have journalists who also are scientifically literate, or at least literate enough to ask the right questions.

Anyone who is even halfway literate, scientifically speaking, asks to see evidence for any position in an argument. Such evidence is not always easily available, nor is it necessarily easy to interpret. The more controversial the topic, the less likely that any member of the general public can get access to all the information needed for making a scientifically valid decision. Abortion is an excellent example of such an issue, partly because it has been the most volatile hot-button issue in American politics ever since the *Roe vs. Wade* decision was made, but mostly because there is no clear cut way to decide the issue on the basis of observation. Every piece of evidence one might use to claim that a first trimester fetus is not a human being is subject to debate because of someone's definition of "human being." Even fertilization involves so many discrete and identifiable steps that when considered from a purely scientific perspective, the word "fertilization" comes to mean a whole suite of events, any one of which could be chosen by someone to define the term. Thus the common pro-life stance that a fertilized ovum is a person and should have all rights accorded other people under the constitution, is itself open for discussion on strictly biological grounds.

So scientifically, the phrase "human life begins at conception" is about like saying "any great meal begins with the choice of lettuce." Both are true, of course, but a great meal could also begin with choice of the salad dressing, the steak, the soup, the wine, or the guest list. Depending on who's the potential cook, the great meal could easily begin with the *idea* that maybe this cook should invite some people over. Thus considering "fertilization" to be a legal defining moment is somewhat akin to buying the head of lettuce, tossing it in the fridge, then telling your friends what a great meal you served a crowd of guests who, to a person, were Renaissance people with much to talk about. Or, for that matter, with some cooks, such as the Food Network stars, for example, just thinking about inviting some friends in for dinner would be completely analogous to defining human-ness as the point of "fertilization." For these culinary experts in our analogy, deciding nah, I guess I won't invite them in after all, would be the equivalent of contraception or an abortion, depending on how close they'd come to actually making a salad.

My point in the above discussion is that when a process is so complex, and involves as many distinct and identifiable steps as human fertilization—and I'm talking about only that short time during which a sperm successfully encounters an ovum and haploid nuclei subsequently fuse—then taking that process as a starting point for anything is scientifically unsupportable. We cannot decide the time, *scientifically*, at which an idea becomes a human being by addressing the process of fertilization. We can very easily decide the time, *politically*, at which an idea becomes a human being by collapsing all the events of fertilization into its "final" result, which is not really final at all but just one more step in a long process of growth and differentiation.

As is the case with so many battles in the culture wars, emotions, simplification, and desire trump observation and rationality; what we *know* does not win over what we *want*, what we *believe,* or what we can convince others to believe. Most of the time, this victory of mind over matter, or desire over reality, does not have very serious consequences. It's only when someone starts a shooting war because of desire and beliefs that the consequences of being scientifically illiterate become a little more serious than they would be otherwise.

Perhaps a more clear-cut example of an issue involving science literacy is needed. Let's say the United States Food and Drug Administration is evaluating a new drug for the treatment of indiscretion. Let's call this drug Levonorgestrel, or a "morning-after pill." Keep in mind that the word "Levonorgestrel" is a proper noun designation for some chemical substance, whereas the phrase "morning-after pill" infers indiscretion, thus questionable moral behavior, so that the word and the phrase can easily be linked politically. There have been about 1600 scientific papers published on Levonorgestrel since the late 1970s, virtually all of them by scientists seeking information on its effectiveness and safety as a post-sex contraceptive.

Among the most significant of these papers is one entitled "Levonorgestrel-only dosing strategies for emergency contraception," published in the journal *Pharmacotherapy* in 2007 (vol. 27, pages 278-284) by Laura Hansen, Joseph Saseen, and Stephanie Teal of the University of Colorado. These scientists reviewed all the pertinent scientific literature published globally between 1967 and 2006, almost forty years worth of study, including work on results from either a single 1.5 mg dose or two 0.75 mg doses within 12-24 hours of unprotected sex, and found absolutely no evidence for either adverse side effects or difference in efficacy, depending on dosing regimen. See the *Boston Globe* editorial for November 20, 2005, and related news stories from a variety of sources. When a history of early Third Millennium medicine is written, it will be noted that once George W. Bush left the office of American president, his successor, Barack Obama, asked for a review of the science surrounding Levonorgestrel and consequently the drug was made available, without prescription, to 17-year olds.

Of course I chose the Levonorgestrel case for its literary and social interest; the vast majority of us love sex, stories about sex, and questions of morality associated with sex, and we all have fairly strong opinions about who should be having sex with whom and when. But science literacy is an issue in many areas beyond sex and medicine. Most of these issues are not very consequential when taken singly, but considered collectively, they can characterize a mind-set typical of a community, a state, or a nation. Thomas Frank, journalist and *Wall Street Journal* columnist, describes an outstanding example of such population-level scientific illiteracy in his best-selling book *What's the Matter with Kansas? How Conservatives Won the Heart of America* (2004). Despite all kinds of evidence to the contrary, Kansans seem to consistently vote against their own vested economic interest; Frank's only credible explanation is that the state simply has been taken over by a conservative ideology that is so severe it blinds the majority of voters to any rational and objective analysis of their own situation.

Frank's book formalizes our impressions of scientific illiteracy at the state level through use of data, that is, a rather scientific approach. He gives us economic figures that nobody argues with (observations, records, neutral information), then proceeds to express his amazement at Kansans' apparent willful ignorance relative to this information. Frank was preceded by Barbara Tuchman, who used a similar narrative strategy in her book *The March of Folly: From Troy to Vietnam* (1984), again asking the question: why do nations behave in ways that are clearly counter to their own vested interests?

Individuals often seem to have the answer to this question; populations rarely do. Thus we have the big take-home lesson for elected officials: *if an action is in the public's best interest, and everyone can plainly see that this course of action is in the public's best interests, yet the public is determined not to carry out the action, then an elected official's job is to convince the public to behave otherwise, not to do what the public wants.* If that sentence seems long and convoluted, then try reading it again a few times. The lesson is not a very difficult one to understand; it can be exceedingly difficult to apply, however, sort of like stopping a runaway train, although in this case the train is made of willful ignorance and outright stupidity.

Kansas again provides us with an excellent example surrounding a fairly important science-related issue, one that unlike abortion is actually fairly resolvable, namely, public health. The details of Kansas' behavior, carried out by its elected officials, can be found in the state's public records, but are summarized by C. J. Janovy (yes, a relative), in Kansas City's alternative newspaper, *The Pitch*. The main players in this little on-going drama are described, in *The Pitch*, as follows: "one is a vice president at a Fortune 500 company, one is a pediatrician, others are medical-center CEOs, two are high-level university administrators, and another is a professor of management at Harvard Business School."

This group is the Kansas Health Policy Authority, and its stated mission is (from its web site): *KHPA shall develop and maintain a coordinated health policy agenda that combines the effective purchasing and administration of health care with promotion oriented public health strategies.* In recent years, KHPA, aided by its 266 employees, has presented a long list of health care items to the Kansas Legislature, only to have the vast majority of them rejected.

The KHPA provides a partial list of these rejected items on its web site (to quote):

(1) Posting health-care cost information on the Internet;

(2) Collecting physical-fitness information on Kansas students;

(3) Encouraging cafeterias in schools and state buildings to serve healthier foods;

(4) Establishing a program for cancer-screening and dental care for pregnant women on Medicaid;

(5) A pilot program to determine whether small businesses would benefit from wellness programs in the workplace.

Given this list of rejections, none of the others having to do with tobacco, for example a $0.50/pack cigarette tax or a state-wide ban on smoking in public buildings, had a chance. What we have in Kansas is a body of elected officials clearly acting against the vested interests of those who elected them. The kindest thing that can be said for these elected representatives is that they are scientifically illiterate, ignorant to the point of being unable to act in accordance with relatively simple data sets. The next kindest thing that can be said about them is that they are really dumb. The kindness quotient falls off pretty steeply after "scientifically illiterate," "ignorant," and "dumb."

Kansas' ongoing political conflict over creationism and evolution in the public school curriculum is not quite such a blatant example of scientific illiteracy at work against the public interest as that of the KHPA recommendations, but it nevertheless illustrates even better the definition of "scientific literacy." Why is the political conflict over teaching of evolution such an excellent example? The answer is very simple: because it illustrates so clearly the triumph of belief over evidence. The vast and overwhelming majority of this evidence for evolution has little or no relationship to human beings; indeed, most of it concerns marine invertebrates and fruit flies. Kansas voters' illiteracy is revealed by their preoccupation with *humans*, a single primate species. But then as pointed out elsewhere in this book, both science and religion are human activities, so it should surprise no one that political conflict over evolution should focus on human origins.

The evidence for a human origin from pre-human primates is strong, compelling, and multifaceted, being assembled from diverse sources such as the fossil record, comparative anatomy and development, and research using modern molecular technology to discern genetic makeup. In a scientifically literate nation there would be no argument over where humans came from, only arguments over what it means to be a human and how we preserve our human-ness in the face of numerous and powerful dehumanizing forces (war, debt, discrimination, poverty, etc.) Then scientific literacy would be only a minor factor in the discussions, whereas philosophy, religious beliefs (= prevailing myths), and behaviors contributing to, or detracting from, societies' abilities to sustain themselves, would be the topics. Maybe if we were elephants or whales, our beliefs and ideas would not be so important, but we don't know enough about the minds (brains) of elephants and whales to decide this question. What we do know, however, is that ideas, beliefs, and perceptions are of enormous importance to us because we are humans, thus are legitimate topics for serious study and discussion.

Why are beliefs, ideas, and perceptions so important to humans, and not so important to elephants and whales? The answer to this question is found in our anatomy, from our hands to our vocal chords. We don't really do anything with these organs that is terribly different from what our nearest relatives do, but we do these things in much more elaborate, diverse, and dangerous ways. Chimps communicate with sounds and facial expressions; humans write operas and make Oscar-winning movies. Chimps strip leaves from a twig to make a termite-extractor, then poke that twig down into a termite mound, retrieve it, and lick off the insects; we build nuclear missiles, as well as submarines and supersonic aircraft to deliver them. Elephants exhibit powerful emotional attachment to their young, caressing them, protecting them, and teaching them how to be elephants; we buy video games, iPods, and cell phones for ours, and these toys end up doing quite a bit of teaching, whether that teaching is intended or not. But most of all, we talk and write.

Every year during March and April, about half a million sandhill cranes, *Grus canadensis* show up between the towns of Grand Island and Kearney, in the central Platte River valley of Nebraska. Concurrently, there arrive thousands of tourists, packing motels and restaurants, heading out long before daybreak to freeze their rear ends off in a blind for close-up views and photo ops of these magnificent birds. The experience is primeval and exhilarating; sandhill cranes are large birds, and along the Platte River many are engaged in mating rituals—"dancing"—with leaps and flapping wings; others slam the cold ground with their beaks, feasting on corn left over from last fall's harvest. When disturbed, they lift off *en masse*; in the still, cold, morning air, punctuated only occasionally by a distant bawling cow, the pounding of their wings sends concussions through your ears. High above, from a circling flock comes the guttural bugling of a thousand birds, a sound amplified and given its unique timbre by a long windpipe winding through their breastbone. Sandhill cranes have been on Earth for about 30 million years. Their unique calls have been on Earth for 30 million years because sandhill cranes make them. Once sandhill cranes are gone, these sounds also will be gone, as gone as whatever vocalizations were made by dinosaurs.

Shingebis and the North Wind, or some version of it, is a story from Native American mythology. In this story, Shingebis defies winter and remains in the north, eventually inviting North Wind into his dwelling for a wrestling match, which Shingebis wins. The myth takes different forms, depending on the tribes that told it. In some versions, for example, Shingebis is a diving duck; in others, he's a person. The story was evidently used as a teaching device: behave like Shingebis—providing for winter, seeking alternate ways to survive, defying the forces that seek to defeat you—and you will prosper (win the wrestling match), or at least spring will eventually come. We don't know how long the Shingebis myth existed on Earth prior to its being recorded in print form; we can guess that the time is somewhere between 10,000 and 50,000 years, that is, about the length of time modern humans have been making timeless art exemplified by paintings such as those on the walls of Lascaux and Altamira.

We don't know what stories were told by Cro-Magnons around their fires, but any species that makes timeless art can be assumed, quite reasonably, to also tell timeless stories. Evidence also is rapidly accumulating from a wide variety of sources that Nendertals, once considered brutes, were indeed far more human than many of are willing to admit. The molecular biologists have shown that some current human populations have Neandertal genes, clear evidence of interbreeding. The same scientists have demonstrated that Neandertals had the gene clusters related to language, and paleontologists have contributed data on mouth and tongue structure (see http://sjohn30.tripod.com/id1.html and the primary literature citations provided). Relatively recent anthropological research reveals use of pigments and natural materials to create decorations and, evidently, jewelry. These folks were not dumb, uncultured, grunting animals; they were real humans. Given Neandertals' geographic distribution—mainly northern Eurasia—it's not inconceivable that the Shingebis story originated in their caves a hundred thousand years ago.

Once it found its way into print, however, the story of Shingebis spread like wildfire far beyond the relatively modern, in anthropological terms, Ojibwe wigwams. This story, probably told thousands of years ago in a language long gone from earth, to children gathered in some kind of a shelter against a howling November wind ripping across what is now Nebraska, can, in the opening years of the Third Millennium, be plucked from my shelf of childhood literature, saved from my parents house decades ago, and read again, in a warm, brick home heated by methane gas, while outside the howling November wind once again rips across what is now Nebraska.

There is nothing particularly unique about the tale of Shingebis and the North Wind; it just happened to be a handy example when I was writing this chapter. What is startling about this example is that unlike the bugling of sandhill cranes, the Shingebis myth can readily be replaced with literally millions of stories, about virtually any subject, and teaching an equal number of lessons, some of value, some detrimental, but all suddenly timeless, made so by repetition in media concocted by humans, media that extend these human utterings far into realms unavailable to whales, chimps, and our Cro-Magnon ancestors.

Why, in a chapter about the meaning of scientific literacy would I digress into subjects such as sandhill crane vocalizations and Native American mythology? The answer to this question is very simple and straightforward: *to illustrate the enormous power of words, power to spread through human populations, power to influence people far beyond those who did the writing or speaking, power to assume many forms, power to evolve and adapt to whatever culture gets infected, and power to alter our behavior.*

The only things a sandhill crane vocalization can influence are the behavior of another sandhill crane and the relatively few humans who respond emotionally to it and have either the political or financial clout to buy a nature preserve. Human vocalizations, combined with sequences of symbols, made using various kinds of technology, can send nations into war. Sandhill cranes have been bugling into the wind for thirty million years. Humans have been talking to one another, insofar as we suspect, for at least 50,000 years. When sandhill cranes are extinct, their sound will also disappear from Earth. Humans, however, leave their traces in writing, art, music, mathematical equations, and technology; Neandertals may be extinct, but at least one of them conveyed the idea of respect for their dead by placing flowers around a body, flowers that left pollen to be found millennia later, an act revealed by science, then interpreted in accordance with contemporary beliefs, that is, the equivalent of words.

I believe that the leading question of our time is: *what is a human being*? The answer, summarized, is: a very large, very destructive, exceedingly intelligent, creative, and handy primate that is quite capable of living, at least temporarily (in evolutionary terms), in fantasy worlds built from words and pictures. What usually gets left behind, once these primates cross over into their imaginary worlds, is their scientific literacy, and along with it, recognition of their evolutionary history, their genetic heritage, and the natural forces that ultimately will govern their lives. We are, ourselves, our species, living the tired old two-edged sword metaphor. Those gifts bestowed upon us by our genetic makeup have carried us, along one edge of our existence space, beyond the chimpanzee to Mozart and Picasso, and along the other edge to Hiroshima and Nagasaki.

The Mozart-Picasso edge leads to products of the imagination, fantasy, stories that stir us but have no necessary connection to our needs for water, food, shelter, and mates. The Hiroshima-Nagasaki edge leads to unimagined destruction, obliteration, or, alternatively, a redefinition, albeit an unflattering one, of our human-ness. Like all swords, this metaphorical one also comes to a point, the sharpened, piercing, tip being that moment when ultimate fantasy meets ultimate destruction. In historical terms, "moment" can easily mean "year," "decade," or "century," but whatever its extent, it's that period when the tip of our sword emerges from the metaphorical state into a tangible one, we pick up a newspaper, turn on the TV, check our Internet news, increasingly on a smart phone, and once again, for what seems like the umpteenth time, discover someone has killed someone else in the name of God.

15. Why is Scientific Literacy of Such Importance?

> ... I wonder whether we may not envisage modernity rather as an attitude than as a period of history. And by "attitude" I mean a mode of relating to contemporary reality...
>
> —Michael Foucault (*What is Enlightenment?*)

Literacy in general, not just scientific, is of profound importance to any civilized society, or, for that matter, to any civilization that is to sustain itself in a particular environment. Of course "literacy" can refer to any domain of information, so that in truly primeval societies such as African Pygmies or the western New Guinea highlands Ndani, the ability to read and interpret signs in the forest is just as important as an ability to read street signs or graffiti marking gang territory boundaries would be to an urban American. But scientific literacy is a special kind of literacy because it addresses our relationships with the natural world on a grand scale, and these relationships with nature are crucial ones because Earth is the only planet known to support human life, and we are humans.

Now, having said that, I admit that vast numbers of people believe that this planet is doomed to obliteration, that Earth is only a temporary home for our bodies, that our spirits will live for Eternity in some far off place, and so whatever actions we take here and now are not really very important in the long term. History suggests, however, and fairly strongly, that such thinking makes truly bad foreign and economic policy.

Widespread scientific literacy is crucial to the welfare of any so-called developed nation for several reasons. First, most such nations are heavily armed, and in the Third Millennium, armaments are fairly sophisticated machines. That is, they are built using technology derived from our understanding of substances and forces present throughout the universe and built in accordance with various scientific principles. We can believe in Heaven but faith alone cannot direct a missile to its correct target; instead of faith, we need computers, software, sensors, explosives, propellant mixtures, transportation, electronic communications, and highly trained people, all products of a scientific enterprise. If we are at war, science allows us to aim our weapons at the enemy instead of at ourselves. And if we view war, and the potential for war, as a major economic engine, as we obviously do in the United States, then basic science to support technological development is crucial to a large segment of our economy.

Scientific literacy also is vital to a developed nation because so many of our public policies, especially ones having economic impacts, are linked to management of natural phenomena. Good examples of this relationship include water allocation, crop subsidies, energy resource development and utilization, natural disaster preparedness, the provision of health care, flood plain designations, and zoning. There may be public debate over "environmental issues," but in the end Mother Nature will decide how much rain to deliver and when to deliver it, how much corn can be produced on an acre of Iowa farm land, and whether to bash New Orleans into oblivion or break San Francisco off the country and dump it into the Pacific Ocean. So "debate over environmental issues" really translates into a contest between what we know and understand about the way nature works and what we want to have happen. In other words, scientific literacy shapes the contest between reality and desire.

This contest between reality and desire is perhaps the most important reason of all for a nation's citizens to be, on the average, scientifically literate. Scientists have a certain mindset, one that is governed by evidence, observation, and technology, and in which interpretations or conclusions are always subject to modification based on additional information. In the vast majority of cases, this scientist's approach to his or her profession carries over into everyday life outside the laboratory. Scientists certainly are not alone in exhibiting this particular type of behavior; artists, attorneys, and physicians, indeed virtually all of us, tend to view the world through lenses shaped by our professions. But we need to remember the fundamental nature of science: an exploration of the universe using falsifiable assertions as the primary working tool, assertions that are developed within the context of a general explanatory theory.

This basic nature of the scientific enterprise generates some rules about evidence used to support assertions. Put bluntly, the scientific mindset demands falsifiable assertions and observations that will test those assertions. Scientists typically heap scorn on unfalsifiable assertions, good examples of which can be found daily in American political discourse and indeed throughout American domestic policy of the Third Millennium. Scientists are equally scornful of assertions for which the supporting evidence is exceedingly flimsy, borderline unattainable, or subject to severe sampling flaws. Some such assertions are so burdened with ideological baggage that studies to test them, while technically possible, are not always politically possible. Again, our public political discourse provides ample illustrations of such assertions. Here are a few familiar ones:

(1) Abstinence-only sex education in public schools will significantly reduce sexual activity among teenage children, unwanted pregnancy, the incidence of sexually transmitted disease, and abortion.

(2) A combination of standardized testing and threatened punishment for low performance on such tests will significantly improve the levels of math and science literacy among

American school children, especially the most disadvantaged ones.

(3) Reducing taxes for the wealthier Americans will improve the economic status of all Americans.

(4) Some kind of a national health care program will bring economic ruin to the United States.

(5) Prescription drugs purchased in Canada are a public health hazard.

(6) Elimination of prayer in public schools leads to moral decay of the nation.

(7) Hollywood is eroding American moral fiber with its never-ending supply of sex and violence.

This list could be longer, and with a little bit of effort, any American could add to it just by reading the newspaper or listening to the radio. Thus we are besieged with assertions that seem to be congruent with our internal logic yet to the scientific mind fail for all the above mentioned reasons. A good example of such an assertion might be: "If we 'teach' abstinence then teens will be abstinent." Inadequacy of data, which reflects mostly an inability to actually obtain relevant data, probably tops the list of reasons for scientific scorn. Assertions that cannot be tested because we can't actually design the studies and get the appropriate numbers can be highly effective political weapons, but these weapons tend to be used on members of the society that develops them instead of on enemies, perceived or real. We turn our arguments upon ourselves and the fact that they cannot be scientifically evaluated means they never go away.

As an example of an untestable assertion, consider the abstinence education assertion mentioned above. To test it, we would need several experimental groups that were carefully matched economically and demographically, and several control groups not taught anything about sex, just to start a truly legitimate scientific study. Modern studies involving humans all require approval by oversight committees, usually ones associated with medical schools, and such approval involves informed consent waivers that in turn require either adult status or parental signatures.

Imagine some scientist coming into a PTA meeting to inform assembled parents of middle school children about this study and its design. In essence, you'd be telling these parents: we're going to teach abstinence to some of your kids but not others. Then we'll measure sexual activity. You now have an explanation why this assertion about the effects of abstinence education is essentially untestable. After-the-fact surveys, however, provide a sort of test, and this sample assertion (= hypothesis) is generally conceded to be rejected. In other words, abstinence-only sex education classes don't have any observable long term effect on teenage pregnancy rates. Consult the following web site for data:

http://www.advocatesforyouth.org/publications/stateevaluations/index.htm).

In general, survey data usually are at least somewhat indicative of attitudes and resulting social change, but tend to be highly variable in "quality." That is, such data are subject to bias arising from ignorance, improperly phrased questions, inadequate sampling strategies, and population traits hidden from people conducting the survey. These potential failings support a professional polling industry, the Gallup International organization being a prime example. Survey design professionals try to write questions that are self-validating, reveal specific attributes, and minimize emotional impact.

In other words, if you are contacted by someone being paid to collect survey data, and if you are willing to answer the questions, you are very likely to be asked as many as twenty, and sometimes up to fifty, questions about how likely you are to exhibit some kind of behavior, with responses limited to phrases such as "very likely," "somewhat likely," "somewhat unlikely," and "not likely." The behavior can range from voting for a certain candidate to buying a certain service or product. Among the questions, however, will probably be some internally-validating ones. Thus if you answer a particular question with "very likely," you will probably also answer the internal validation question the same way, assuming you are telling the truth about your experiences, attitudes, or behaviors.

A scientifically literate citizen understands, or at least appreciates, the variability inherent in study design, the range of reliability of data on social issues, and the ideology built in to assertions by politicians. A scientifically literate citizen always asks first about evidence that a particular assertion will be, or has been, true. One excellent example of an assertion that probably should have been subject to close scrutiny typically reserved for peer reviews of scientific studies was Public Law 107-110, otherwise known as the 2001 No Child Left Behind Act (NCLB). The assertion (= testable hypothesis) was this one: "A combination of standardized testing and threatened punishment for low performance on such tests will significantly improve the levels of math and science literacy among American school children, especially the most disadvantaged ones."

Any teacher knows that this testable hypothesis is actually a formula for derailing an educational system that might be doing about as well as could be expected, given the resources, level of parental involvement, and economic status of students' families. Thus if you're threatened with punishment because of low standardized test scores, you teach to the test. If you're required by law to report performance by student category, then you divert human resources into reporting. As any scientist could have predicted, the main result of NCLB legislation was a booming statistics industry and a generation of students, and their teachers, who are far more concerned with *the* answer to *a particular* question than with acquisition of transferable skills such as reading, writing well, and understanding quantitative issues in the public realm (taxing practices, interest rates, public indebtedness, real costs of military action, etc.).

Scientific literacy, like visual and other forms of literacy, influences a citizen's views about, and actions relative to, public policy and behavior of elected officials. Scientifically literate people tend to adopt positions, on public policy, that reflect an understanding of all factual information available. Sometimes religious or other beliefs shape opinions, but scientifically literate people recognize this influence and understand when they are acting out of personal belief, even if that belief overrides rationality. The national debate over stem cell research exemplifies this situation beautifully, and the State of Nebraska, of all places, with its elected University of Nebraska Board of Regents and heavily-Catholic constituency, provides the clearest illustration of the role played by science, or non-science, in this political conflict. Because of its clarity and simplicity, the Nebraska case is worth exploring in detail.

The University of Nebraska Board of Regents has eight voting members, elected by districts and serving six-year terms. A complete description of this board and its duties can be found on the web site: http://nebraska.edu/board/. The institution they are elected to govern includes units ranging from a vocational agriculture campus in Curtis, Nebraska, to the University of Nebraska Medical Center (UNMC) in Omaha, the latter a sprawling, expanding, and well-funded behemoth a few blocks away from another well-funded behemoth, Creighton Medical Center, the latter supported by the Catholic Church. Omaha is at least 65% Catholic and Lincoln, where the main—that is, football-playing—campus (UNL) resides, has one of the nation's most belligerently conservative bishops.

The fall, 2008, elections saw a battle for a Regents' position between Earl Scudder, a middle-aged, highly knowledgeable, broadly-educated, and experienced attorney, former president of the UNL Parents Association, and Tim Clare, a young, very Catholic, attorney, son of Pat Clare, orthopedist and Husker football team physician. Clare's campaign focused on two issues: in-state tuition for children of illegal immigrants, which he played up far beyond its financial importance, and stem cell research at the university's Medical Center. Naturally he was against both.

Both issues, it turns out, were covered by existing laws, regulations, and policies, but Clare evidently was convinced, and probably correctly so, that the University could, if it so desired, add to those laws and regulation. The assertion that illegal immigrants were costing the state significant amounts of money because their children were given resident tuition at our colleges and universities was simply wrong, although it was played up as an outrageous insult to American sovereignty, maintained by a duped Board of Regents. The simple truth was that state law more than adequately covered this uncommon situation, and if some poor kid actually made it to the university, managed to graduate, obtain citizenship, and become a productive, tax-paying, citizen, then both the state and the nation would benefit.

The stem-cell research issue had long been resolved by a set of Regents-approved guidelines that were relatively restrictive and completely consistent with National Institutes of Health regulations, giving sperm and egg donors total control over the fate of their frozen embryos. As far as Nebraskans in general were concerned, the issue of whether and the University of Nebraska Medical Center (UNMC) personnel should conduct research on stem cells was a forgotten issue. In the political realm, however, the phrase "stem cell research" has functioned very well as code for "killing babies." Moreover, in the public mind, the term "research," often conjures up images of Federal funds spent on meaningless arcane projects such as the evolution of flies. (Do a Google® search on "golden fleece award" for an historical account of such projects.) What [now] Regent Clare seemed to be promising the electorate was use of the power of his office to invalidate years of serious faculty and administrative work, eventually approved by a conservative Board, to tightly regulate politically sensitive areas of university activity. In other words, he reminded us of what we'd already accomplished, and rekindled a settled issue, one settled by extreme professionalism on the part of all concerned, for his own political gain.

It's a little difficult to determine why people run for the office of University of Nebraska Regent, aside from the fact that they get football tickets in the press box with luxury, enclosed, seating, and trips to bowl games. The Board position is not very functional as a springboard for higher office, and once elected, an individual discovers very quickly that the work-to-glory ratio is much higher than originally envisioned and that one's capacity to effect major changes in society is severely limited. Past Regents have been relatively diverse, although mostly conservative, some astonishingly so, and usually finish their terms feeling like they've performed a public service and that's enough of that, regardless of the football tickets and bowl trips. For the record, as of this writing, the UNMC stem cell research policies and practices, and university system policies regarding resident tuition, along with the state law governing resident tuition, remain unchanged. Regent Clare has his tickets to athletic events, and as of the day I am writing this paragraph, a pretty good chance of attending a bowl game somewhere warm in the middle of the Nebraska winter.

What does this example, involving a relatively minor election in a relatively unpopulated state have to tell us about the nation's scientific literacy and its importance? The answer is: quite a bit, although I'm sure with a little bit of searching, anyone could find a dozen or more equally useful scenarios as reported in the media. The illegal immigrant and instate tuition issue probably is the more glaring case of public illiteracy in action. State law provides resident public college and university tuition for any immigrant who has lived in Nebraska for three years, graduated from a state high school, and declared his/her intent to become an American citizen. The number of such individuals applying for admission to Nebraska public colleges and universities is relatively small, although in states with similar laws, for example, California, the number is likely much higher, and because out-of-state tuition in California is so high, the financial impact is substantial.

What's missing in this highly simplified discussion is the intangible benefit of a young person graduating from high school with an education adequate to support an application for admission to college, then following that graduation with actual attendance at college. Presumably, subsequent graduation from a college or university prepares one for a job, or a career, with a regular (taxable) paycheck, a desire to purchase real estate, and a rather extensive list of life-long expenditures of the type that support a healthy economy: groceries, clothes, automobiles, services of all kinds, insurance, travel, and, ultimately, college tuition for children. This social/economic trajectory is, in the vast majority of cases, accompanied by successful application for citizenship.

Is the alternative to this "American Dream" scenario a life of drugs and crime? Not necessarily, but then it doesn't take very many drug pushers and street gangs to cancel out the economic benefits of one well-employed, law-abiding, tax-paying citizen. A scientifically literate person, unafraid of numbers, might easily conclude that there are plenty of cases in which public policy that offends an individual, and seems quite counter to the tenets of Christian morality, does in fact work for the common good. Malcolm Gladwell, best-selling author of books addressing social issues, is an excellent source of examples. For one such case, see "Million-dollar Murray: Why Problems like Homelessness may be Easier to Solve than to Manage," (*The New Yorker,* Feb. 13, 2006). The bottom line is that the law enforcement costs of dealing with this one individual far outweighed the cost of a treatment and supervision program. Similar data exist for needle exchange programs (do a Google® search using "needle exchange program").

In the end, the main question for a scientifically literate person is: which is more important to a civilized society, solution to a costly social problem or adherence to my personal beliefs about individual behavior and responsibility?

Ideally, these two factors match and public policy that reflect my personal code of moral behavior, supported by religious documents and practices, does in fact solve the costly social problem. A scientifically literate person, however, says "good luck; don't bet the family farm that this will happen, and certainly don't bet the farm that your personal religious beliefs make the best public policy for a highly complex, technology-dependent, society."

16. Why are Politicians so Scientifically Illiterate?

> *A United States policy that could find no other option, he suggested, was one of "indolent short-term expediency."*
> —Barbara Tuchman (*Stillwell and the American Experience in China 1911-1945*)

Elected officials in general are scientifically illiterate for two reasons: first, they often are either businessmen or –women, or attorneys, and nether one of these professions requires or encourages scientific literacy; and, second, real scientists and serious teachers, those who are most likely to be quite literate, typically have neither the taste for, nor the resources to seek, elected public office. In a pluralistic society such as ours, both reasons are fairly legitimate.

Later on in this chapter I'll come back to the business community and its scientific literacy because this subject is a rich one to use in exploring the interrelationships between science, technology, national security, and economic health. The legal profession, however, has little or no reason to be scientifically literate except in cases involving modern forensics or industries that are heavily dependent on technology. Lawyers and businessmen can, however, and often do, hire their scientific literacy in the form of consultants and expert witnesses, either of which may be quite literate, but neither of which is particularly constrained by the unwritten laws of real science. In other words, consultants and expert witnesses don't necessarily *do* science; instead, they are skilled *consumers* and *users* of science.

In a previous chapter I defined "scientific literacy" and explored its several applications. In this chapter I'm actually going to address the problems of science education, but not necessarily what we typically think of as "science education" in the public school sense. Instead, I will focus on what we might call "deep" education, that is, the kind that changes the way we view the world. For example, we might claim that having a vague sense of what a molecule is, and perhaps even being able to define the term, counts as being at least somewhat scientifically literate. After all, you have a word and an idea in your mind and you're not completely baffled when you hear the word spoken on television or read it in the newspaper. Furthermore, you might actually be able to use this word in a complete sentence, such as "I wonder whether the molecules in these pills will make me sick if I swallow them with good Irish whiskey," or "I wonder whether some of the molecules in that bag of lawn chemicals will kill my cat." These particular sentences reveal an incipient scientific-type curiosity, whereas the sentence "Don't bother me with all that talk about molecules, just give me something to cure this headache," although also a complete sentence, nevertheless expresses a naïveté typical of the scientifically illiterate.

The deep education is revealed, however, when you incorporate the idea of a molecule into your daily decision making, regardless of whether the decisions are simple easy ones (whether to put sugar or artificial sweetener on your cereal) or more long term and difficult ones (whether to stop taking your prescribed medicine because of a newspaper report on associated side effects in a small number of cases). In the first instance, you feel comfortable making the simple decision because you also have a third choice, namely neither, and you don't know anyone who's actually been hurt by either sugar or artificial sweetener, at least in single doses. Your molecular decision may be influenced by your weight on any particular day, by your weight on previous days, on your desired weight, or on the feeling of having achieved a goal relative to weight control. Although the decision to use sugar or substitute is a pretty trivial one in scientific terms, when that decision becomes part of an overall engagement with matters of diet and weight control, especially for sound and healthy reasons, then the decision indicates a fairly sophisticated engagement with biology as a science, albeit at a highly personal level. And, if you actually read labels on food products and understand most of what they contain, you're well on your way toward becoming scientifically literate.

The second instance, namely, the decision to stop taking a prescription drug, is more troublesome because you really don't have much control over many of the factors that went into your possession of this supply of molecules. You did not write the prescription; your doctor wrote it based on observations that you might know but probably don't completely understand. You don't have any information beyond what's written in the newspaper about the serious side effects cases or from various web sites, some of them provided by the pharmaceutical industry and others provided by kooks. In the best of all worlds you quit taking the medicine and don't notice much difference in your health because the medicine wasn't having any dramatic effects anyway (this actually was the case with me and a drug prescribed for joint pain). In the worst of all worlds you start worrying about the potential side effects and can't seem to get a straight answer from your doctor or HMO. So it becomes a relief when the company that manufactures this drug pulls it from the market. The deep decision has been made for you.

The decisions that I've called "deep" are ones that involve both a breadth of scientific knowledge and a propensity, derived from an understanding of science as a way of knowing, to evaluate evidence supporting an assertion and to think in comparative terms. Deep decisions accept the fundamental nature of science, namely its dependence on observations, the independence of those observations from your desires or beliefs, and the fact that to be relevant, observations must function to test a testable assertion. Such decisions also accept the idea of probability and the fact of statistical variation instead of demanding certainty. In the case of the sugar substitute decision, you may well have shown a high level of scientific literacy if you engaged in all the label reading and diet design activities intended to keep you healthy and actually knew why you were doing these acts. In addition, such literacy probably primed you to acquire further scientific knowledge and understanding if needed, e.g., when faced with a significant environmental issue affecting your property values.

Furthermore, if you're convinced that your dietary awareness, exercise, and label reading keeps your weight and cholesterol under control, then you're sort of a walking experiment but with a sample size of one and no control group with which to compare yourself. Nevertheless, you have a testable assertion regarding your own body and through your activity based on scientific literacy you are testing that assertion about weight and blood chemistry, and probably also self-esteem. Regardless of the sample size and lack of control group, you are making decisions that affect yourself and perhaps others, such as family members, using knowledge about the natural world, and using that knowledge in a rational way consistent with scientific practice.

The alternate version of this assertion, that if you consume large quantities of certain kinds of molecules you will become heavier and less healthy (as revealed by a bathroom scale and the blood work at annual physical exam time), is also well within your power to test, but your decision not to test it is an example of one based on a meaningful relationship between desire and nature. In other words, nature will allow you to fulfill your desire, provided you interact with nature in a way suggested by scientific knowledge about how nature—your body—works.

As indicated in a previous chapter, testable assertions are the hallmark of science, and I'll expand on this scientific property within the context of political action later in this chapter. But for the moment, we should remember that in the political arena, assertions are testable only within an historical framework. In other words, politics is an historical discipline with its own rules of evidence that may not match those of proximal or normal science, i.e., the kind of science that does experiments with material amenable to experimentation. Within the realm of history, you can't really do "experiments," as we properly define the term; you can only assess the validity of some assertion by looking back on what actually happened when you acted as if that assertion was true.

There is no better example of this kind of historical assertion testing than the Iraq war that began with the invasion of that nation by a group of other nations, led mostly by the United States, in 2003. The assertion that Iraqis were developing, or had, and intended to use "weapons of mass destruction," the assertion that Iraqis would quickly adopt an American-style democracy once their dictator was overthrown, and the assertion that Iraq would be a business-friendly working environment shortly after hostilities ceased, all were tested and shown to be false. But unlike a real experiment, say involving bacterial metabolism, you can't go back and start over with Iraq.

The vast majority of all politicians rely on public approval to sustain their employment. In addition, once in office, the trappings of power can become quite seductive. These two facets of political life are among the main reasons that politicians are so scientifically illiterate, or at least act as if they are. Nevertheless, most if not all positions occupied by politicians also involve major responsibilities, compliance with various laws, ceremonial activities, and nowadays, public scrutiny of religious beliefs and behaviors demonstrating "faith." Nobody who professes to be an atheist should be so stupid as to spend money running for public office in the United States of America, no matter how lowly that office might be or how qualified the individual. Elected membership on the Lancaster County, Nebraska, Weed Control Authority comes immediately to mind; no self-proclaimed secular humanists need apply. Thus politicians are scientifically illiterate, or act as if they are, because the demands of public office, the need for public approval, and the constant scrutiny of their faith-based behavior, all job-related phenomena that work to make such literacy a liability instead of the asset it should be.

Besides the factors of responsibility, approval, and scrutiny, it is also important to remember that mobs want answers and solutions, not questions and problems, from their leaders. In general, science tends to produce more questions and problems than answers and solutions. This tendency derives from the fundamental nature of science as an activity. Elsewhere in this book I use the metaphor of an island of understanding in a sea of ignorance to explain why science produces more problems than solutions. Remember that as an island grows in size (increase in understanding), its shoreline (the boundary between understanding and ignorance) also grows. All the questions and problems lie along this boundary. In addition, to continue with the metaphor, the larger an island gets, the more geographically diverse it tends to become. If that geographic diversity involves mountains, then we have a high perch from which to observe the sea of ignorance. Routinely such observation shows that sea to be much larger than we imagined when we were only down on our hands and knees in the sand studying nature at the [metaphorical] shore.

The familiar case of New Orleans vs. Hurricane Katrina beautifully illustrates all these points about breadth of knowledge, comparative thinking, observations, history, and the basic properties of science. Breadth of knowledge is perhaps the most important factor that should have been considered in the political decisions involving the Mississippi Delta ecology. Thus a broadly educated politician would never simply ask how much money an ecological project—for example, a system of levees and an artificial river (the New Orleans shipping channel)—costs, or how much money the public is willing to spend on such a project. Instead, as a minimum, a broadly educated politician considers history, socio-economic conditions, the probability of disaster, the quality of expertise consulted, whether or not that expertise is in agreement with other expertise from diverse sources, the nature of observations, the process of analysis, and whether the process itself has obvious flaws or internal contradictions. In other words, to really assess the adequacy of New Orleans levees, one would have to study the Mississippi Delta using approaches that would be quite familiar to any evolutionary biologist.

Research over the past half century, i.e., activity increasing both the size of our island of understanding and the length of its shoreline boundary with the sea of ignorance, clearly revealed (produced) more questions and problems about the Mississippi Delta region than answers and solutions. Such research involved new technologies such as satellite imagery, geographic information system software, and socio-economic analysis, as well as experience derived from study of the Achafalaya River and its basin using more conventional methods—measurement of stream flow, sedimenttation and erosion rates, pressures on diversion dams and gates, etc.

Over the years, the scientific community came to realize that the initial problem and its solution, namely, keeping water out of New Orleans by building levees, was actually only a small part of a much larger problem, specifically, long term management of the interrelationship between a nation's economy and one of the world's largest rivers. This kind of collective activity, in which a truly massive ecosystem is the primary player at the center of a highly integrated, far-reaching, transportation and financial network, does not lend itself to governance by mobs that want answers and solutions, not questions and problems, from their leaders. Instead, this kind of system requires almost Jeffersonian dignity, patience, foresight, and breadth, traits that don't survive well in our Third Millennium media-driven electioneering environment.

Such a broad education, and its use in a public arena, is therefore a lot, indeed probably too much, to ask of any modern politician. But then, of course, it is the job of any newspaper reporter half-way qualified for his or her job to ask the right questions of elected officials in order to reveal their breadth of knowledge, in situations involving natural phenolmena, or, in the best of all worlds, to inspire those politicians to acquire knowledge, wisdom, and some decent honest advisers who are not just sycophants. Sadly, perhaps for reasons that are deeply embedded in the human DNA, as a general rule we are not patient with careful analysis, complex interactions between elements of nature, varying degrees of probability, and leaders who are honest about the chances that disaster will befall us. Instead, we seem to admire leaders who are strong advocates of actions based on our beliefs and desires, who inspire us to be courageous, and who tend to simplify a complex universe down to issues and explanations we can understand. And leaders who can convince us we are in danger, and seem to be fighting that danger in an obvious way, are the ones we seem to admire the most. None of this typical interaction between a population and its chosen leaders promotes scientific literacy or honesty about the relationship between nature and people.

I do not claim that scientists, because they are scientists, are more honest or broadly educated than politicians. In the realm of science, however, the honesty system operates much more strongly and rapidly than in the realm of politics, mainly because this system typically involves anonymous review of scientific work before that work is made public, and it does not involve public decision-making. If you are doing experiments on the sex life of some tiny worm, and try to publish your results, i.e., make them public, then some well-educated scientist will scrutinize your methods, including your experimental design, your statistical analysis, your rationale for doing the project in the first place, your interpretations of the results, the extent to which you have taken existing knowledge into account, and even the quality of your writing. All this review does not necessarily make you an honest person, but it does tend to pick up flaws in your thinking and mistakes in your actions. But if you go to a cocktail party filled with attorneys and elected city officials, the main question you are likely to be asked about this research is: "Why is this kind of stuff important?" The question really means: "Why are you wasting time and money, maybe even tax money, on this kind of activity, and why do you seem to be so interested in sex?"

There may be two thousand good reasons why you are studying the sex life of obscure worms, but these reasons probably involve the fundamental nature of science itself. These worms could, potentially, become a model system for the study of hormone action at the cellular level, thus serving to help explain developmental anomalies in humans, livestock, and companion animals. The worms might be extraordinarily beautiful creatures under the microscope, thus quite attractive to students who in turn could easily become internationally renowned scholars studying some global human affliction but who remember fondly their carefree undergrad days back in the lab when all they had to talk about was worm sex. The worms' reproductive biology could easily shed light on the origin of sex itself, or the evolution of pheromones, both subjects of enormous interest to the scientific community. Pheromone action, as you might suspect, also could be of substantial interest to the cosmetics industry. When a scientist hears that another scientist is studying the sex life of obscure worms, then all of the possibilities mentioned in this paragraph usually come to mind because scientists typically understand how science itself works on a grand scale. Politicians, however, like their constituencies, rarely get past the issues of time, money (especially tax money), and sex, although sometimes, if not often, there is a hidden disdain for people who would spend their lives studying microscopic creatures with no immediate economic importance.

In our example of the worms, politicians' focus on time, money, sex, and utility is not necessarily stupid, evil, or dangerous, although it has the potential for being all three. In the previous paragraph, I've actually revealed all the reasons why in order to remain economically competitive in a technologically competitive world, any nation needs to have a strong, healthy, broad, and active scientific enterprise. *Flourishing scientific activity, sustained largely by curiosity about the natural world, breeds scientists, models, new ways of studying nature, and new applications of existing technology.* In other words, it is the *human resources* that are of prime importance to a highly developed nation, not the discoveries themselves. Given enough human resources engaged in research, techniques for studying heretofore mysterious aspects of nature will be developed and the discoveries will be made. Furthermore, breadth of research interest tends to produce transferable technologies, a critical factor in sustaining a technology-based economy.

The laser (light amplification by stimulated emission) is perhaps the best example of this phenomenon from 20th Century science in the United States. A brief history of this technological development can be found on the Bell Labs web site (www.bell-labs.com/history/laser/), but in essence, two scientists—Arthur L. Schawlow and Charles H. Townes—developed the technology from research that began in the 1940s. Schawlow was a researcher at Bell Labs, and Townes was a consultant to the Bell Labs research enterprise. These scientists' primary interest at the time was molecular structure, and the laser was intended to be a device to help them pursue their research in this area.

The commercial development of this technology, along with its rapid spread throughout almost all aspects of modern American life, can be traced to the publication of a paper entitled *Infrared and Optical Masers* (in *Physical Review*, volume 112, pages 1940-1949, published December 15, 1958). You can read this original piece of science simply by doing a Google® search using the title—*Infrared and Optical Masers*—as your search term. Now we have laser pointers in the classroom, laser surgery in the hospital, laser scanning in the grocery story, etc. Although the laser may be the most easily understood example of transferable technology, our daily lives are filled with other cases. And, of course, science feeds on itself in this regard, with practicing scientists always looking for new applications for new and existing technologies.

Another history lesson—actually a rule of human resource development—that politicians typically fail to understand is the following: *Artists often spring quickly, even spontaneously, out of a population, but scientists do not.* Technological advances and economically important innovations might periodically emerge out of the realm of basic science, but the "realm of basic science" requires a vastly different cultural milieu than does the intellectual soup that spawns artists. Any nation that does not outright suppress or punish artists will end up with a good supply of them, but to be economically competitive in the Third Millennium, nations need lots of healthy, authentic, curiosity-directed, scientists and such individuals are not guaranteed to arise, and become legitimate scientists, by virtue of their own two hands and a paintbrush.

Nowadays, science needs physical facilities, computational power, technology, ready access to information on a global scale, time, and patience. To be economically competetive in the Third Millennium, a nation does not need a bunch of ignorant elected officials, afraid of science, afraid of the word "evolution," and afraid of anything that seems to support immoral behavior. A nation needs, instead, a bunch of courageous and intellectually honest people who have the interpersonal and verbal skills to help educate its citizens on scientific matters, and especially on the link between economic health and a valid understanding of how nature operates.

17. Is "Evolution" Dangerous?

> . . . by an unyielding dedication to the war, he had come to personify the American endeavor in Vietnam. He had exemplified it in his illusions, in his good intentions gone awry, in his pride, in his will to win.
> —Neil Sheehan (*A Bright Shining Lie*)

The answer to this question is very simple: *no*. The term "evolution" refers to both a scientific phenomenon, that is, a theory and the data to support it, and a natural phenomenon, namely, a process. Theories and the research that spawns them are not dangerous or evil, but natural events and processes can be dangerous depending on what they are and on their extent and proximity. Tsunamis, tornadoes, volcanoes, and earthquakes are natural phenomenon, and manifestations of geophysical processes, that are dangerous if one is close to them, but that have been happening on Earth since long before there were humans to worry about them. Theories are completely neutral, no matter how close you get to them.

Remember that science is a human activity and scientific phenomena are observed, constructed, or developed by humans. Without people on Earth, there would be no theories but there would still be plenty of snow, rain, volcanic eruptions, fires, floods, landslides, continental drift, and evolution. It's what humans do with their constructions that produce danger and evil. For example, consider the old familiar saying: guns don't kill people; people kill people. We could say the same thing about atomic theory: particle physics *per se* never killed anyone but nuclear weapons sure have.

Evolution the theory is seen by some as dangerous because it does not invoke a supernatural power with ultimate judgment over human fate, thus does not provide a reason for human beings to be kind, considerate, honest, and faithful to one another. In the modern political debate, this freedom from ultimate judgment by a supernatural power is presumed by some to lead humans inexorably toward dancing, murder, narcotics use, homosexuality, abortion, adultery, and other pursuits such as video games and rap music, all of which are considered immoral by at least a few of us. In other words, as the logic goes, if we are not threatened with eternal damnation, and if we do not accept all the strictures imposed by a particular organization built on belief, then we are free to behave in all kinds of anti-social and deviant ways and this freedom leads inexorably toward, perhaps even *causes*, such behavior. I am not claiming that this line of thinking is correct, only that its logic is internally consistent, provided you begin with the premise that God's final judgment is the necessary, although not always sufficient, force that produces human behavior seen as beneficial to and acceptable within human societies.

On the other hand, there is plenty of evidence from the historical record that all kinds of immoral, irrational, and downright stupid and self-destructive behavior occurred among and between humans long before Charles Darwin was ever born. Indeed, the Bible is, if nothing else, one long narrative on the post-Eden degeneracy of our species, the various wrathful punishments our behaviors brought down upon our heads, and the advice or admonishments of various prophets. The Bible was codified, in more or less its present form, about 1400 years before Charles Darwin was born and about 1450 years before *The Origin of Species* was published. Most theologians agree that the Bible, especially the Old Testament, rests firmly on historical events of the previous several millennia, although such history should probably be considered highly modified by various agendas and narrative techniques used by those who recorded it. For an introduction to this issue see David Chidester's *Christianity:*

A Global History (Harper Collins, 2000) or the long article and especially the online conversation regarding disputes, at the familiar and communally edited website

http://en.wikipedia.org/wiki/The_Bible_and_history).

As an alternative to these two recommendations, you can always visit a large public library and search its catalog using the term "Bible history." You'll get enough responses, including some excellent scholarship, to last a lifetime, should you decide to pursue them. (The University of Nebraska-Lincoln library returned 2,646 hits, virtually all of them also containing extensive bibliographies.) The Chidester volume and two books by Karen Armstrong (*The History of God*, Knopf, 1993, and *The Battle for God*, Knopf, 2000) are excellent summaries of the information in those 2,646 hits. Based on history alone, one has to say that evolution is a decided late comer as a stimulus for immoral behavior, if not an outright pretender to that dubious honor. Until *The Origin of Species*, behavior perceived as immoral was blamed on Satan, who evidently was directing people to do things that God forbade. Only after Darwin was "evolution"—the ideas, theory, and observations and their interpretations—blamed for immorality, or perhaps assumed the role of Satan as a potential *cause* of immorality.

The modern perception of "evolution" as the cause of immorality can be traced back to a number of sources, but especially to the movement called "social Darwinism," although we must also blame lay persons' misinterpretation of Darwin's theory, and especially of the boundary conditions under which this theory applies to living organisms. Again, our friend Wikipedia provides an adequate introduction to this movement, but I encourage all users of that wonderful online resource to scroll down to the references and primary sources bibliography at the end, not only of the material on social Darwinism, but of the entries on any topic on which information is sought. In summary, the tenets of social Darwinism consider competition to be the primary factor driving success and failure of human societies and nations. Thus in the minds of many men, "survival of the fittest" easily translates into financial and political success, often, if not especially, at the expense of others, typically with little or no regard for the "others," regardless of who they may be.

However, and this is a rather significant "however," remember that Darwinian success, or Darwinian fitness, involves only reproduction, not other measures of "success," as the term is routinely defined by humans. Wealth and political power are not measures of Darwinian success; procreation is. Breeders win, no matter how ugly, destitute, uneducated, heavily armed, dangerous, and extremist they may be, or we believe them to be. Indeed, those who are most fit today, in the Darwinian sense, are not the CEOs of AIG and General Motors, the owners of professional football teams, or real estate magnates, but the destitute millions, if not hundreds of millions, who occupy the tropics and sub-tropics. And regardless of how destitute and fecund they may be, many of them also are heavily armed and quite capable of causing mayhem in a variety of creative ways. Darwinism indeed works on societies, just not in the ways our former robber barons would have had us believe.

On the other hand, it is not clear that anyone in Victorian England or Industrial Revolution era United States actively proclaimed social Darwinism as the guiding theory or justification for their political, military, or business decisions. Throughout the past several millennia, the desire for power and money, especially on the part of males, has directed such decisions far more than has any social theory. Indeed, social theories tend to be developed by nerds and assessed, if not applied, in retrospect when history is interpretted.

But like religious beliefs, theories can become dangerous in the hands of humans when political leaders justify actions based on them regardless of whether those leaders understand the theories or not, or even realize they are behaving in accordance with some general explanation of how the universe operates. This danger is especially real when so-called "theories" evolve into ideologies. Thus one does not have to understand evolution, Darwin's ideas, the evidence he used, genetics, molecular biology, phylogenetic algorithms, alternative modes of speciation, or even the simple equations predicting population growth (equations learned by most 9th graders in the developed world) to declare that "we are a Christian nation" (not actually a theory, but an untested, and probably untestable, hypothesis) and subsequently ban marriages between homosexuals or go to war in an Islamic land halfway around the world.

Regardless of whether it is testable or not, the phrase "we are a Christian nation" can function like a theory, that is, like a general explanation of some natural phenomenon, and in doing so guide and/or justify actions. It could be argued, perhaps successfully, that in the opening years of the 21st Century, "we are a Christian nation"—the quasi-theory—is far more dangerous to the United States, in the long run, than evolution, the real theory. The reason for this danger is evident from the historical record: populations of *Homo sapiens*, the so-called wise primate, routinely fight and kill one another over religion, but they rarely if ever do so over science. Perhaps it is time to ask whether the not-so-wise primates behave in similar ways, or whether they simply fight over food, territory, and potential mates.

Evolutionary theory predicts that if we are products of primate evolution, as a truly preponderance of evidence suggests, then we should be able to find examples of behaviors exhibited by non-human primates, some of our closest relatives, that reflect some of our own. Alternatively, such behaviors could be considered unique to humans, that is, a product, or byproduct, of whatever forces made us humans and did not contribute to the making of apes. In evolutionary parlance, traits are either *primitive* or *derived*; that is, they are either shared by all related forms within a group (= primitive), or they are novelties, arising by some set of events (= derived). Biological scientists commonly, and with good reason, consider such novelties, typically characteristic of a species, to be the product of evolution (the process). So immoral behavior is either derived (= uniquely human), or primitive (= shared by all descendents of an ape/human ancestor that exhibited such behavior, at least in recognizable form).

Do thievery, adultery, infanticide, murder, and disrespect for one's parents—all behaviors proscribed by The Ten Commandments—occur among monkeys and apes? The answer is: most certainly, and furthermore, such behaviors are very well documented. Immorality is primitive, not derived, but when bonobos do it, nobody calls it immorality; instead they call it . . . well . . . "behavior," or perhaps "behavior characteristic of the species," or, depending on who "they" is, "fun and a little bit titillating" (bonobos spend a lot of time on sex). If the East African Rift Valley was Eden, then chimps, gorillas, and bonobos were chowing down on apples from the Tree of Knowledge, and were well aware of the potential roles played with and by their genitals, for a very long time before Adam and Eve came along.

Although theories themselves are not particularly dangerous, other kinds of human constructions made only of words, ideas, and symbols can seem to mimic theories, in that they serve as general explanations for observations about nature, and such explanations can be a problem for societies, regardless of whether they are testable or not. The above example—"We are a Christian nation."—is but one of many. But it's not really clear whether problems caused by such quasi-theories are particularly dangerous to the societies that face them. In the opening years of the Third Millennium, undeveloped quasi-theories, for example, socially conservative ideologies such as those espoused by the American religious right, including certain mega-churches and televangelists, have become both a shield and burden.

The shield is multi-functional and can protect those who hold true to the ideologies from having to accept as appropriate a long list of behaviors seen as somewhat threatening to a uniform, structured, and tightly controlled society. Good examples of such anti-uniformity behavior include appreciation for abstract art and classical music such as Liszt's *Mephisto Waltzes* (1859-1885), reading magazines such as *The New Yorker* and *Harper's*, ignoring your neighbor's sexual orientation, and playing golf on Sunday morning. Extreme conservatives may consider such behaviors dangerous to our nation, but in fact, they're pretty inconsequential ones, especially when compared to wars of choice (do a Google® search using "Iraq war" as the search term).

The burden imposed by quasi-theories and ideologies is that they typically do not take into account the biological nature of human beings. In other words, if you firmly believe that *Homo sapiens* is a special creation, made in the image of God, then it's a struggle (burden) to also believe that *H. sapiens* is a mammal with certain ingrained behaviors, extreme variability, extraordinary intelligence and free will. Such a burden can easily prevent those who hold to the ideologies from addressing real social problems with rational solutions. An excellent and recent example of this principle at work can be found in the article by Margaret Talbot entitled "Red Sex, Blue Sex" (*The New Yorker*, November 3, 2008).

This story deals with teen pregnancy, especially unmarried teen pregnancy, a rather obvious and real social problem. Evangelicals, it turns out, with their uncompromising moralistic ideals, are experiencing a boom in unwed teen motherhood, almost to the point of seeing it as a right of passage among high school seniors. More liberal, secular, and worldly, families, however, with a more realistic view of youth behavior and the value of birth control, seem to be having a lowered teen pregnancy rate. In this case, the evangelicals' ideology—inconsistent with all we know about human biology—essentially creates a social problem. This example is but one of many, if not dozens, that could easily be cited, and any citizen with access to a library should be able to find others quickly.

Remember that in the minds of scientists, theories are general explanations for natural phenomena, explanations that not only are testable, but also have been tested repeatedly, by various individuals doing research by accepted methods that in other cases have yielded practical applications. For example, molecular techniques used to test evolutionary hypotheses are exactly the same ones used in criminal investigations. This link between hypothesis testing by scientists (often an arcane activity, depending on the subject) and practical application is an important one for informed citizens to understand and appreciate. Once a technique, such as determination of nucleotide sequences in DNA, is commonly available, then the choice of problems to solve using that technique can become quite long and diverse.

Using DNA as our example, we could address some question about form in living organisms, for example, as whether similarity in structure reveals true evolutionary relationships. Alternatively, we could apply that technique to a criminal case, using DNA recovered from semen to determine identity of a rapist. In general, American society appreciates the use of technology in criminal investigations; it is not always obvious that this same society appreciates what research on relatedness of plants, animals, fungi, and microbes—using the same technology—tells us about the history of life on Earth. Nevertheless, the latter research often constitutes a test of hypotheses derived from theories, in this case, evolutionary theory.

Real theories give rise testable hypotheses, which actually are corollaries. Evolutionary theory is our example, although gravitational theory or quantum theory could just as easily suffice. If they are strong enough, and if they survive serious and repeated testing, generating a multitude of testable hypotheses that occupy the time and talents of many scientists, then theories can take on a life of their own. In other words, they live in the literature independent of people, but they can also live in the collective mind, and in the culture, in a manner analogous to some organism. Having made their "escape" from the realms of their scientist owners, theories are then "out there" for anyone to use, but this freedom also allows theories to intrude into human affairs in sometimes unwelcome ways.

Darwinian natural selection, a.k.a. "evolution," is our prime example. Free of the abstract boundaries created by professional science, it becomes a weapon, a factor in political campaigns, a criterion by which we judge fellow humans, and justification for attitudes and beliefs. Furthermore, "evolution" evolves (!) into a higher form of social phenolmenon, namely, a standard by which morality, in the highest, that is, Biblical, sense, is assessed. Some post-millennial American citizens, with free access to educational opportunities of a bewildering variety, will indeed declare you unworthy of, in fact ineligible for, "salvation" if you "believe in evolution."

In this case, in which "evolution" is used as a criterion to separate the saved from those condemned, it is not the theory itself that is dangerous, but the use of it as a weapon. Such use also demonstrates rather clearly that people are quite capable of using almost anything as a weapon in their various conflicts with one another, including shunning, which is the use of *nothing* (no recognition that another human being exists) as a weapon. So no, evolution is not dangerous. In fact, it's not really even appropriate to ask whether evolution is dangerous. Evolution is an explanation for life's diversity, and to whatever extent it can become dangerous, that is the extent human beings can use the ideas associated with evolution as weapons.

Thus we might ask: in the United States, against whom are these weapons being used? The answer is again reasonably easy to supply: against Americans, against the nation, and against our children. Those engaged in creationist politics actually are seeking to promote, spread, and institutionalize scientific illiteracy in a heavily armed, heavily indebted, military hero-worshipping, nation that also is heavily dependent on scientific literacy and losing ground daily, in the economic wars, against more rational and secular nations such as China.

18. What is a Human Being?

> *A theory that explains eating by man's digestive system cannot explain also why different societies obtain and prepare food in different ways, or why some have food taboos and others do not.*
>
> —Peter Farb (*Man's Rise to Civilization*)

This question—*What is a human being?*—is the leading question of our time and it can be answered in about seven billion ways, i.e., in as many ways as there are people on Earth. At two of many extremes are the answers that lead to the so-called creation-evolution controversy: (1) A human being is a special creation made by God in His image; or (2) A human being is a highly evolved ape, a species whose biology is dictated by bipedalism, an opposable thumb, and a hypertrophied cerebral cortex.

There is little to be gained by denying that humans are animals, specifically mammals, whose physiology and genetics are in most ways very similar to those of other mammals, especially pigs and great apes (chimpanzees, gorillas, and orangutans). Indeed, anyone who denies this similarity reveals a deep and inexcusable ignorance about life on Earth, or else a political agenda in which ignorance is a weapon. The genetic and physiological similarity between people and other animal species is undeniable; it is an observed fact. Medical research, with all its discoveries and health-related technology, rests firmly on this similarity.

But humans are not *exactly* identical to other species, either genetically or physiologically, and certainly not intellectually. The genetic and physiological differences are well known and taken into account during trials of new medical procedures. Our brains, however, set us an almost indescribable distance apart from other species, not so much in terms of neuron function or motor responses to stimuli, but in terms of that emergent property we call the "mind." Thus we have the capacity to create worlds, indeed whole universes, out of only thoughts, ideas, words, and pictures.

We have no idea whether other species can perform this remarkable feat, and we really don't have the tools—intellectual or otherwise—to even conduct the studies necessary to determine whether orangutans can make fantasy worlds where they then go hang out when life in the trees gets boring or stressful. But if we have a single defining trait, it is our ability to make something out of nothing, then go live where our minds have taken us, devoting considerable time, energy, and money to such imaginary travel, and in the process acting as if our constructions are truly natural phenomena.

This trait—that of being able to build a fantasy house then live in it—is, of course, of short time economic value only under certain environmental conditions. Those conditions require that we be free of the responsibility for finding food, water, and shelter in the wilderness, and for avoiding real predators. Civilization accomplishes the feat of generating just such conditions, at least for large numbers of us. Thus the mind's most wonderful and outrageous creations can be considered products of socio-economic systems characterized by the exploitation of natural resources, the division of labor, and the generation of leisure time.

Lest we equate such systems with post-WWII northern hemisphere developed nations, we must remember that for centuries, human beings, including those in so-called primitive societies, have produced absolutely wondrous art, literature, and music, along with highly effective technology designed specifically for the natural environments occupied by those societies. The more "primitive" literature may be relatively inaccessible except to those already educated in appropriate languages and history, and the music may grate on our rap-trained ears, but the art is on display around the world, most dramatically, perhaps, in the relatively new Musée du Quai Branly in Paris. This striking, $333,000,000, architectural statement houses thousands of pieces of non-Western art, and the collection, preservation, and display of such anthropologically significant material is the museum's primary mission. That is, its goal is to answer the leading question of our time: *What is a human being?*

Not everyone agrees that the artifacts in the Musée du Quai Branly are "art," not because the pieces are not beautyful, or visually imposing, or fail to send some subliminal message to a viewer, but because unlike, for example, American Abstract Expressionist paintings, such constructions come laden with culture-specific baggage and their designs are deeply intertwined with tribal traditions, oral history, and mythology. Wandering the Branly galleries, a single question surfaces repeatedly: What was going through their *minds* when they made that . . . [fill in your own item, e.g., mask, shawl, stool, etc.]? Yes, indeed, what was going through the *minds* of people making tangible objects, in a particular *way*, at any place on Earth and in any historical period for what must have been a reason that went well beyond utility but was intimately tied to that utility?

Thus we have one significant and general answer to this leading question of our time—*what is a human being*? A human being is an animal for which utility is a necessary feature of constructions, but quickly becomes a secondary property, i.e., *necessary but not sufficient*, and thus only a starting place, upon which elaborations are placed, elaborations that reveal the maker as human. A chimpanzee does not spend time decorating a twig used to extract termites from a mound, but a human being carves decorations, perhaps of importance in terms of oral history, on a knife handle, then on the shoulder blade of some bison that previously was dinner.

A chimpanzee may pick up a limb and use it to bash a competitor, but the animal doesn't save that particular limb, then scratch intricate designs upon it. On the other hand, a human being makes golf clubs that regardless of the underlying technology must look a certain way, resting on the ground between one's feet, and engraves memorable scenes on weapons, in the process turning them into collectibles. Few would argue that guns are, for example, in and of themselves, art; equally few would deny that the diversity of weaponry available to any law abiding American citizen goes far, indeed, quite exceedingly far, beyond that required to kill squirrels, ducks, quail, deer, or other people.

A human being therefore is an animal that not only makes art, music, and literature, none of which is of immediate utilitarian value but also turns strictly utilitarian items into decoration, a personalized form of art that in turn diversifies, almost as if having been invented, it is then free to manifest a phylogeny. Furthermore, the reverse also is true, and seemingly useless items always have the potential for becoming transformed into something of lasting economic value. For example, a Rossini overture can, if you stretch your definition of utility a little bit, serve to keep orchestra musicians, stage hands, instrument makers and repairmen, and printers employed, at least at minimal wages, so spread over time, the use of such music is roughly equivalent to the use of some invention—people do things with it, and go to some effort to keep it within the realm of accessible social constructs, so there is money to be made from this work.

When there is money to be earned, then bills can be paid and children educated. Similarly, in utilitarian terms, a Rothko painting is absolutely worthless; nobody ever killed an enemy, fixed a leaky faucet, or took the kids to soccer practice with a Rothko. But museums holding these gloomy abstractions attract thousands of visitors, many of whom pay admission, so money changes hands because someone wants to see a painting made by someone else. People work in security, preservation, the coffee and gift shops, development and fund raising, and education; gift shop items often are crafted by local artisans in distant lands and imported from halfway around the world; airline companies, hotels, and taxi drivers all partake of the Rothko mystique indirectly. This flow of money and jobs surrounding the visual arts exists because of Rothko and the thousands of others who produced art to satisfy some inner message emanating from their *minds*.

The same claim can be made about our fascination with history and the natural world. Globally, various museums preserve our evidence for what the world was and is really like, what we as a species have built, used, and admired. Chimpanzees do not build museums; only humans purposefully resurrect then consume cultures, reflect on the results of our behavior over millennia, and build astronomical observatories. Chichen Itza and Machu Picchu are products of the human mind, especially that part of the mind that wonders about our relationship to the natural world and seeks to define humanity's place in it.

The American Museum of Natural History, the Smithsonian Institution, and the Muséum National d'Histoire Naturelle are prime examples of our humanity, as much a revelation of our minds at work as the Tate Modern in London and the Museum of Modern Art in New York City. The list of museums in Paris alone is pages long, and although Paris is probably the finest example of humanity's preoccupation with itself and the place it lives, all major cities have an impressive number of similar institutions and the collection/reflection habit extends even to small towns out in the American heartland. Pick out any place in Kansas and do a Google® search; chances are excellent you'll find a chamber of commerce tab on the town's web site, and along with it, a list of historical and artistic attractions. Try the Museum of the Fur Trade in Chadron, Nebraska, or the Roller Skating Museum and the Museum of Germans from Russia in that state's capital city Lincoln. Oklahoma City is awash in such blatant displays of our human-ness: The National Cowboy Hall of Fame and Western Heritage Museum, the National

Softball Hall of Fame, and the 45th Infantry Division Museum, to name but three of a dozen.

Insofar as we know, chimpanzees don't do genealogies, either, buying software to assist in the exploration of one's own identify, spending hours at the computer tracking down relatives, sending cheek swabs to commercial DNA analysis labs, taking trips back to the homeland, and loading up the mini-van for a long haul to Salt Lake City for personal searches through the Mormon Church records. Orangutans don't collect their maternal grandmother's China, their paternal grandmother's silver, or display pictures of their infants in a treetop office. Gorillas don't put correspondence in museum archives, build Presidential libraries, or write biographies. Non-human animals stare at themselves in mirrors, and we have some evidence that they are aware of their own identity, especially in the case of great apes and elephants, but none of these species has the borderline pathological curiosity about themselves displayed by humans, at least insofar as we know or can determine by existing research methods.

The museums and amateur genealogists of unremarkable communities such as those in Lincoln, Nebraska, and Oklahoma City, Oklahoma, are not particularly unusual. Instead, they are outright pedestrian examples of our narcissism, a fact that reveals the extent of it. We *want* to know about ourselves, we *want* to be reminded of what we have done and where we have been, both literally and figuratively, we *want* others to know about us, and once we're old enough to master a keyboard, we swarm to Facebook.com and start broadcasting messages about who we are and who we want to be.

Unfortunately, however, this seemingly intense curiosity about other members of our species, including those quite unlike ourselves, is matched by an equally intense repulsion of other humans, especially those who appear on the surface to be quite different from us. The Musée du Quai Branly is predicated on the former; racism, ancient ethnic hatreds, and the Iraq civil war of 2005-? are sustained by the latter. A leading question for theorists in the area of social and developmental psychology might be: how are these two urges related? A corollary question could easily be: how do societies make the transition from one to another, or at least reconcile the conflict in a way that does not wreak havoc?

These questions are not merely mind games for academics. As *Homo sapiens* spreads its destructive power across the globe, consuming tropical forests at the rate of 50 acres a minute, stripping the oceans of life (over two million tons of fish and other seafood a year eaten in the USA alone), sucking its rivers and fossil water supplies dry, burning fossil sunlight like there was an infinite supply (which there is not, and everyone knows it), and killing one another by the tens of thousands a year, the bases for our conflicts and group behaviors should be of truly global concern.

In an utopian world, we would come to understand why we destroy, fight, kill, and act against our own species' interests, then apply that understanding to resolve conflicts and limit our impact on the planet. We do not live in utopia. In response to the leading question of our time—*What is a human being?*—our best answer so far is: a primate that is probably too smart, or at least too self-conscious, for its own good. And even those traits would not spell our downfall were it not for that damned opposable thumb.

The elegance and sophistication of art, music, and literature produced by our species defy description, and their manifestations, made possible largely by available technology, surround us daily. As a handy example, video games are, to some, a symbol of youthful degeneracy, if not an outright dangerous cultural weapon that "Hollywood" has placed in the hands of our innocent children. Nothing could be further from the truth. First, video games generally are designed and produced in places far from Hollywood, although the use of that name as a synonym for "the entertainment industry" is reasonably close to being justified. Second, they are a highly visible demonstration that science and technology typically have multiple applications, and artists have never been reluctant to use technology in ways its developers never intended or imagined. Third, they represent exploitation of science and technology for commercial purposes, hardly a behavior foreign to Americans or, for that matter, to modern humans in general.

Indeed, in this country we have an expectation that technology will have commercial applications, and most large American universities have a "technology transfer" office whose *responsibility* it is to make money and political hay off of faculty ideas and research. Finally, video games are little more than stories in which players can become involved as participants; in this last sense, they can be classified in the same general category as whatever tales were told around Cro-Magnon fires by some shaman/artist crawling out of the hole where he'd been painting a wooly mammoth a quarter mile deep in a secret and sacred chamber.

A human being is, in the final analysis, an ape that tells stories, only a fraction of which are true, then acts on the lessons of those stories regardless of their veracity. This habit, and the capacity, are intimately linked to our fantasy worlds, the ones we create then go live in, whether those worlds be the ones of novelists, wannabe professional football team owners, or even elected public officials declaring that some faraway peoples who have never in their wildest dreams ever wished to vote for a city council member will quickly adopt a representative form of government once our bombs stop falling on them.

Furthermore, we see stories in art forms that are not necessarily intended to be narratives, and when those art forms don't have obvious narratives, then we are perfectly capable of finding or creating "back-stories." Indeed, the development of back-story—a story that explains or justifies another story—is now an accepted, if not expected, element of most of our literature regardless of the genre or media. The Rothko paintings in Paris' Tate Modern museum, for example, which in and of themselves "say" little beyond some communication consisting of combination of visual, emotional, and contextual emanations, end up saying plenty when provided with back-story. See Wikipedia® to start. The Internet changed Mark Rothko into easily accessible narrative. Rothko was simply a handy example; there are at least seven billion others that would serve as well.

We have absolutely no evidence whatsoever that non-human species use metaphors and other kinds of symbolism in their communications. These literary devices are simply tools to enrich stories and extend their lessons well beyond the literal narrative; Genesis I is a perfect example. Although we have no evidence that chimps or other non-human creatures use metaphors (e.g., "tree of knowledge"), there is a distinct possibility that they use paralanguage, i.e., words equivalents uttered in particular manners and in various contexts, perhaps combined with posture and facial expression, and thus communicating something quite different from the literal translation. In fact, paralanguage may be the primary form of visual/oral communication between non-human vertebrates as well as among many invertebrate species.

As an illustration of this phenomenon of paralanguage, consider the simple phrase "Happy Birthday to you." Most of us in the United States have said this phrase and sung it many times. We've heard it at our own birthdays, and at restaurants sung in obnoxious clapping ways by the wait staff. Within my own family, the song has evolved into an excuse to communicate with our children in a special way, typically early in the morning before any of them really wants to be awake. My wife's birthday is in August, and since childhood it has always been a special day for her, so I ignore it at my peril. My birthday is in late December, thus quite subordinate to Christmas, which my parents celebrated mainly because they probably felt like they had to instead of wanted to, and so I don't remember my birthday ever being important to anyone, especially me, and it still isn't. I cannot remember a single instance in which someone or some group sang "Happy Birthday" to me before I got married and our children grew up enough to do it. Even if Marilyn Monroe had sung it to me, especially in the way she sang it to John F. Kennedy, I would not have heard any of the words.

"Happy Birthday to you" thus means many things to many people, only one of which is really "Happy Birthday." The same could be said of "I love you," "go to Hell," and probably every other phrase uttered by a human being. Thus as is the case with other of our traits, we do the same things that non-human mammals, and often birds, do, but we do those things in a much richer, more complex, manner, and we use the traits in ways perhaps uniquely human. Our paralanguage comes to include our economic condition, the cars we drive, the bumper stickers on those cars, our clothing, the furniture in our homes, the subjects of our conversations, the magazines we read, cell phone ring tones we download, desktop wallpaper on our laptops . . . this list could easily go on for several pages. Words are the essential vehicle for our communication, but human words really are only the equivalent of a stage; paralanguage is the play.

Language and paralanguage are, of course, inextricably linked to behavior, both that of individuals and that of mobs. And of all our defining traits, there is one expressed in behavior that perhaps characterizes us best of all, serving as a summary of all others. Like most of our other seemingly defining traits, this one also is found among non-human species, but is never manifested to the same extent, at the species level, as ours. And what is this so-called behavioral "trait"? It is the difference between our behavior as individuals and our behavior as a group, so can probably be considered meta-behavior.

Virtually all of our highest achievements, from science, mathematics, and engineering, to art, music, and literature, are mainly an individual's works, although admittedly sometimes those individuals led groups of varying size. Throughout all recorded history virtually all of our most regrettable and despicable acts are the work of societies, or at least of large, well-organized, groups. If anyone needs a clear demonstration of how language and paralanguage can make large groups behave in self-destructive ways, he or she need only study the George W. Bush and Dick Cheney speeches that dragged the United States into that economic black hole known as the second Gulf, or Iraq, War.

The human tragedy, of course, is that these mob actions are so often initiated, and subsequently driven, by individuals in positions of power, those exhibiting "leadership." If our species has a fundamental design flaw it is this willingness to let individuals lead groups astray. The most irreverent among us extends that line of thought to organized religion. The most patient and reflective scientists consider such self-destructive group behavior, inspired by individuals that somehow bubble up into positions of extreme power [over the group], to be little more than an inherited trait suddenly become dangerous in a highly technological world built by the very species that possesses such fossil behavior, a behavior more appropriate for survival of roving *Australopithecus* bands than for modern developed nations finding themselves dragged into a global economy.

All of the above discussion defines us by what we do; i.e., by assuming that our identity as individuals, as states and nations, and as a species, lies in our deeds. Scientists, however, are not bound by such assumptions. Lewis Thomas, for example, in his surprising best-seller book *Lives of a Cell: Notes of a Biology Watcher* (1974), summarized our knowledge of evolutionary biochemistry by pointing out that he personally, and by implication all of us, are products of an ancient invasion in which bacteria, or their ancestors, colonized other cells, then devolved into mitochondria, the structures that convert simple sugars to useable energy for all familiar plants and animals. Thomas was relieved not to have to take responsibility for controlling his mitochondrial chemical reactions in order to read the newspaper.

Richard Dawkins followed Thomas' literary achievement with a now classic work in biology: *The Selfish Gene* (1976), which reduced all Earth's inhabitants to mere carriers of nucleotide sequences, the real players in the epic saga of life on a lonely planet. Dawkins' view is that evolution is best depicted as a contest between blocks of genetic information, and that in the larger scheme of things our bodies, as well as those of all past and present organisms, are simply vehicles by which these genes compete with one another for opportunities to persist and multiply. But Dawkins came full circle with his analysis, perhaps inadvertently, by coining the word "meme," which refers to ideas, innovations, even phrases, that move through human populations in a manner strongly analogous to mutant alleles. And so in Dawkins' final analysis we can be defined as "human" by our ideas, inventions, and language regardless of how selfishly those entities behave, proliferate, come to dominate their realms, or go extinct. A case, maybe even a very strong case, could be made for the assertion that our stories rule us.

The virologists, however, equipped with almost surrealistic molecular technology, tell even a more intriguing, and even startling, story about our genetic makeup: much of the human genome consists of viral sequences, molecular memories, one might say, of ancient and repeated colonization. This discovery it not entirely unexpected, given that science has known for a long time that many if not most organisms contain DNA sequences revealing ancient symbioses, and that genes themselves can and do experience duplication, with gene progeny then moving to different chromosomes, giving rise to "offspring" of their own, and living an evolutionary life that may or may not coincide with that of their vehicles—the larger and more familiar organisms that carry these genes around to work and play. Alien scientists, visiting us with the analytical tools of modern molecular biologists, could easily conclude that Earth was a planet of microbes, some of which had the misfortune to end up in an environment that seemed determined to make itself extinct—namely, us.

Obviously this initial question—*What is a human being?*—is one that can be answered honestly in only one way:

we are what the scientists tell us we are, but that scientific answer reveals a rather wondrous mammal whose deeds, both good and bad, continually exceed our expectations. Has our initial question always been the leading one regardless of when and where it was being asked? Perhaps; based on our limited knowledge of history, especially that of diverse cultures, many of which are now extinct, we might honestly conclude that all the stories, rituals, practices, traditions, and arts of those cultures were focused on defining humanity for a particular group in a particular time, place, and ecological setting. And throughout history, again insofar as our knowledge lets us see into distant realms, the dehumanization that is so often a prelude to violence is predicated upon differences in beliefs, practices, traditions, rituals, and stories.

Thus it is a small wonder that people kill one another so readily over differences in religion; indeed, such behavior, so deeply embedded in our species' genetic makeup, is to be expected, and methods to counter this behavior should be the primary focus of whatever training we provide to potential diplomats. When our founding fathers tried to separate religion—that most powerful means of both awarding and withholding membership in our species—from the state—that most powerful means of conducting sanctioned violence—they recognized just how pervasive and all-consuming is our attempt to answer the leading question of our time and place—*What is a human being?*—or our willingness to use the answer for both good and evil, no matter what the calendar date, the latitude, or the longitude.

19. Are Humans Evolving?

> *Quince, usually Beethoven's most persistent groomer, was never observed even attemptting to groom her father during the six months of his recuperation. Instead, she spent a great deal of time sitting by his side gazing anxiously into his face as if trying to console Beethoven by her presence.*
>
> —Dian Fossey (*Gorillas in the Mist*)

Are humans evolving? To cut to the chase, as they say in the movie business, the answer is "yes." More properly, however, the answer probably should be "yes, and very rapidly." All the scientific evidence we have indicates that the Earth's resources are severely strained, and that poverty, disease, malnutrition, competition for energy, fresh water shortages, and rampant population growth are typical of our species' condition throughout much of the world. All of our scientific evidence also shows clearly that the ethnic (~ genetic) makeup of *Homo sapiens*' population is changing on a global scale. Our species is largely Asian and becoming more so by the year.

So any reasonably well educated biologist would look at the combined data set—resource use, population sizes, demographic makeup of various nations, reproductive rates and contributions to human genetic information pool on Earth—and immediately conclude that our species is rapidly approaching that period of history in which food, water, and energy are likely to limit populations, and in which the most successful ethnic groups are those that produce the most offspring. Over the next hundred to a thousand years, those groups will be the Chinese and Indians.

Scientists can make predictions about populations using demographic pyramids, which are nothing more than bar graphs showing relative numbers of individuals in various age groups. There are population pyramids for many areas of the world, again showing the age/sex makeup of human populations in various regions. The United States Census Bureau web site provides such pyramids for every country in the world, so that if you have a computer, then you have immediate access to the same information used by the scientific community to develop evolutionary predictions (see www.census.gov/ipc/www/idbpyr.html). The main issue regarding humans is whether they must obey the relatively harsh rules that tend to govern the lives of other animals, especially other mammalian species. Regardless of your religious beliefs, there is abundant evidence that as far as Mother Earth is concerned, *Homo sapiens* does not occupy a privileged position among species.

Humans require food, water, shelter, and mates, and they produce offspring and wastes. Human internal physiological mechanisms are similar enough to those of other species, especially pigs, that we can do medical research on non-human mammals and apply the results to people. Our genetic traits are coded for by DNA, just like those of all other organisms. Abundant evidence supports the assertion that we are as tied to Earth, its resources and natural processes, as are any of the other species with which we share this planet. When it comes to food, for example, the only thing that distinguishes you from the park squirrels is your grocery store. When it comes to shelter, you buy a house or rent an apartment, not seek out a hole in some tree or make a nest of leaves high in the branches. But when it comes to finding mates, at least some of our behavior is as entertaining as any squirrel's.

The major differences between people and other mammals are to be found in our brains and in our hands (see also the previous chapter). We are extremely intelligent, and our hands are built in a way that allows us to use that intelligence to in turn make machines of all kinds. To go along with our intelligence, we are extremely curious. Although many other species also exhibit high levels of curiosity, they are not always able to do much about the objects of this apparent fascination. A blue jay, for example, can be exceedingly curious about where peanuts come from, but the bird can't do anything with a peanut other than eat it or hide it. Even our closest relatives the chimpanzees could be almost pathologically curious about the private lives of termites—which they routinely extract from mounds using sticks—but without microscopes and laboratories to use them in, as well as hands built to manipulate those microscopes and the surrounding laboratory glassware, chimps are forever separated from termites except as food items. I have to admit, however, that no human has a clue what chimps might *think* about termites.

Humans have most if not all the equipment necessary to satisfy our curiosity, namely, our hands, brain, and languages. With this equipment we have built a world that just seems quite separated from the same natural reality of chimpanzees, giraffes, and butterflies, and we have done this building mainly to protect ourselves from the elements, provide ourselves with food and water, and most importantly, to satisfy our curiosity. Why is this curiosity satisfaction activity more important than finding food, water, and shelter? The answer is fairly simple: scientific research, driven by curiosity, spawns technology that then provides us with additional power over the environment. "Power over the environment" actually translates into the ability to continue providing ourselves with food, water, shelter, and mates, all quite independent of the local weather or predators. Consequently, our *perception* of the human condition is based almost entirely on a world that we ourselves have constructed: a world of cities, highways, machines, organizations, businesses, governments, political ideologies, art, music, and literature.

On the surface, this world constructed by people just looks like it is independent of planetary resources and processes, but natural disasters such as earthquakes, tsunamis, and Hurricane Katrina remind us that it is not. Our legal documents call such events "acts of God," primarily in an attempt to absolve certain parties of blame for unforeseen and uncontrollable circumstances that can in effect void a contract, but hurricanes and tsunamis, as well as hundred year floods, volcanic eruptions, droughts, hail storms, and tornadoes really are acts of Earth. The existence of petroleum also is an act of Earth, but the consumption of that petroleum is an act of people, as is the conversion of petroleum into human flesh through agriculture. In essence, there is a fixed amount of petroleum on Earth because although it may be replenished by natural processes, that replenishment is so agonizingly slow that it is of no consequences whatsoever to humans. So petroleum is a limited resource that ultimately will dictate how many humans can live on our planet.

The human population is now soaring exponentially, perhaps out of control, and this fact is not a mystery to anyone. For example, the figures below are repeated in some form in virtually all freshman biology texts.

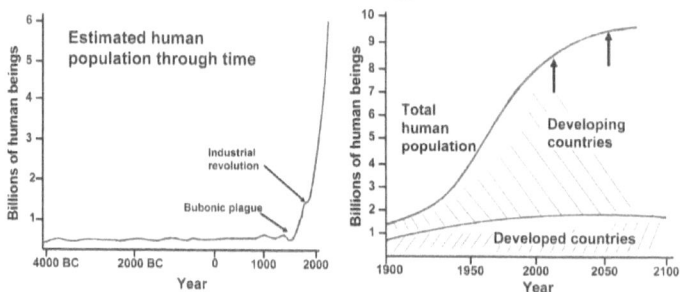

The two arrows on the right hand figure represent the present (2011, as of this writing), and the time when scientists predict that the global population will level off at about nine billion people, that is, at the time current freshman college students are nearing the end of their life expectancies. The two figures together present a classic picture of Darwinian natural selection at work, namely burgeoning population of some species, an eventual limit to the population of that species, and a concurrent changing of the phenotypic makeup of the species' population. Even as we become more numerous, we also are becoming more genetically diverse. Any biologist in his or her right mind would say "yes, *Homo sapiens* is evolving, and at a very rapid, almost visible, rate."

The primary question, of course, is: what resources limit the growth of human populations? Virtually all scientists will agree that land, petroleum, and water are the main ones that come immediately to mind, mainly because we all know that Earth's supply of land suitable for agriculture does not grow exponentially and that petroleum supplies, regardless of how much we find, will eventually run out. Earth might easily make more fossil fuel, but the rate of this process is so much slower than the rate of human reproduction that yes, we will eventually run out of oil. Fifty years is not very long, and a hundred and fifty years is the equivalent of two human life expectancies in the developed world. So as mentioned above, fossil fuel of all kind, but especially petroleum because it is directly linked to agricultural production through planting, harvesting, and transporting grain, is the most important factor that will limit human populations.

Virtually all scientists also agree that there is a fixed amount of water on Earth, and whatever water now exists on the planet simply moves around in its three main forms—liquid, solid, gas—according to a variety of mechanisms. The human body is about 70% water, so quite a bit of water is tied up in human bodies, in fact, about 600,000,000,000 pounds of it, which rounds out to about 72,517,985,612 gallons. Lake Superior is the largest body of fresh water in the United States; its volume is about 12,100 km^3, envisioned as a cube 14.3 miles in all dimensions, which equals 31,794,757,632 cubic feet, or 165,332,739,686 gallons, or about two and a half times as much as is currently tied up in living human bodies.

Lake Baikal, in southern Siberia, is the largest body of fresh water in the world, with a volume of about 23,000 km^3, which equals 816,247,000,000,000 cubic feet, or approximately 6,105,530,000,000,000 gallons, or somewhere around 555,048,000,000,000 people, which scientists estimate is between 5,000 and 6,000 times the carrying capacity of Planet Earth. Obviously there is plenty of fresh water on Earth for the human population. Getting it from Lake Superior or Lake Baikal to where it is needed is quite a different matter.

We use water for many reasons, most or all of which are at least indirectly associated with the building of human tissues: agriculture, manufacturing, generation of electricity, medicine, etc. Agriculture, however, is the big user of water, which means that we are spending lots of water, indeed some estimates are up to 70% of available freshwater, to grow food. Thus we are consuming most of the planet's available water simply to store water, in the form of cytoplasm, in ourselves. For example, it takes approximately 1000 tons of water (~240,000 gallons) to produce a ton of rice, about 450 tons of water (~107,900 gallons) to produce a ton of corn, and 1360 tons of water (~326,098 gallons) to produce a ton of soybeans. Water requirements for wheat are a little more difficult to calculate, mainly because wheat is grown quite differently than corn or rice, but a rough estimate is over 2000 tons of water (~526,000 gallons) per ton of wheat.

Don't bet the farm that these figures are anything other than approximations; I had to do quite a bit of conversion (gallons or bushels to pounds, etc.), but they are based on a wealth of readily available information. Some figures were given in metric tons (=1.1 US tons), so the weight estimates may be off a little bit if my sources did not indicate whether the unit was metric or US standard. For sources, see Internet sites listed in the References chapter, or simply do your own search on Google®, calculator and notepad in hand.

In 2009, the world's corn production was estimated at 817,110,509 million Mt (metric tons), rice at 678,688,289 Mt, wheat at 681,915,838 Mt, and soybeans at 210,900,000 Mt (again, see figures in agricultural information web sites in the References). Thus a single species of primate used

1.96×10^{20} gallons of water, or about 32,000 times the volume of Lake Baikal, to produce the 2009 world corn crop. Given that Lake Baikal has enough water to make about 5,500 times the human carrying capacity for Planet Earth, these figures mean that in 2009, our species, *Homo sapiens*, used water equivalent to 176,000,000 times the human carrying capacity for Earth, *just to grow corn*. Similar calculations for rice, wheat, and soybeans produce equally startling results. And, of course, we have not even considered all the other crop plants currently used by *H. sapiens*, including those with edible roots and leaves.

It is obvious from the above figures that freshwater has the potential for being a rate-limiting resource, not because of its volume, but because of its distribution; Lake Baikal, for example, is simply not available to irrigate sub-Saharan Africa. In addition to its massive role in agriculture, however, water also is a vehicle for the movement of disease-causing organisms. Thus with increasing human population, the current distribution of water-borne infections—cholera, cryptosporidiosis, cyclosporosis, amebiasis, etc.—especially in children, actually is a peek into the future. And what does this future look like, a thousand, two thousand years hence? As is the case with all species, the phenotypes that survive most successfully in times of limited resources are the ones who either adapt to the limitations or compete successfully for whatever resources are available. In times of water stress, the Kalahari Bushmen and Australian Aborigines, at least the ones still living in the ancestral lands, are the most likely survivors because eons ago they adapted to life with severe water restrictions.

Biologists are sometimes portrayed in the media, and by conservative politicians, as doomsday prophets and tree-huggers. The simple truth is that we do not know when doomsday will arrive, but we do know that it *will* arrive, and potentially within the lifetime of our grandchildren, that is, during the time scientists predict that the human population will stabilize at about nine billion. Any respectable scientist should also be willing, and able, to admit the difference between life on Earth and life in Heaven or wherever eternal life is predicted to occur according to various religious documents written since writing was invented six or seven thousand years ago. That reality of life on Earth actually is the reality of a large, exceedingly intelligent, technologically sophisticated, highly destructive primate reproducing like crazy, indeed exponentially, in a realm of fixed resources. Those *Homo sapiens* alive in the next millennium, or the one after, may well be witness to, and indeed participants in, what current scientists call a bottleneck—a massive reduction in numbers of humans followed by evolutionary adaptations among the survivors over the subsequent generations. In other words, not only are humans evolving, and fairly rapidly, but that evolution also could, and probably will, accelerate in the centuries ahead.

Should this evolution be stopped, and if so, why? The question of "should" is not applicable in evolutionary biology. Humans will evolve regardless of whether humans believe they are or should. Organisms of all kinds evolve, period, regardless of whether humans believe they are or should. Thus the truly interesting question is: what will current humans evolve into? Nobody can answer this question, but we can make some predictions that are consistent with current knowledge and observations. Under conditions of environmental stress, genotypes and phenotypes that already cope relatively well with stressful environmental conditions are likely to continue doing relatively well, although "relatively" is a key aspect of this survival. No successful investment banker in New York City believes that peasants in rural China are living relatively well. But any smart investment banker knows, however, that such peasants are living without a whole lot of energy-, resource-, and water-expensive commodities.

The best guess is that over the next several millennia *Homo sapiens* will evolve, and fairly rapidly, into what it is already rapidly evolving into, namely, a smaller, darker, species, subsisting on far less energy, raw materials, and water than citizens of the so-called "developed world" are doing currently. Technology is a major contributor to the suite of characters that defines *H. sapiens,* so an equally interesting question is what kind of technology will this new species have at its disposal? Again, we have no way to answer this question accurately, but we can make a well-educated guess at what it will *not* be, and that is anything that depends to any great extent on petroleum, for example, airplanes, including jet fighters. Technology that ultimately depends on coal, including coal-fired plants that generate electricity, is likely to survive longer than petroleum-dependent technology mainly because the Earth's coal supply has been spared, somewhat, by our ravenous appetite for oil over the past century. Additionally, extraction of coal from the Earth does not require the same level of machine technology as petroleum—think offshore drilling platforms vs. mines in Virginia. But coal also is a fixed resource, namely fossil sunlight, thus will eventually be exhausted.

Homo sapiens has already evolved, and with stunning speed, into a species that is dependent on information technology, which itself is a byproduct of all other kinds of technological developments in areas ranging from metallurgy to machine tool design. So an intriguing question is: can our species give up its information technology and still retain its humanity? And a related question, of more contemporary, and very real, importance is: have we already become parasitized by our technology to the point that we live in a symbiotic relationship? Anyone with even a smattering of training in parasitology would answer "yes, of course!" to this second question, which makes the first one even more intriguing.

What does our technological future look like? Again, nobody can say for sure, but the manufacture of our current technological wonders is highly dependent on various resources. Copper, lead, aluminum, platinum, tin, gold, silver, bromine, silicon, and mercury are among the raw materials needed to build a computer. For battery-operated devices, including cell phones, add nickel, lithium, cobalt, cadmium, and zinc. And, of course, there are plastics. Manufacture of silicon chips is a highly sophisticated business, demonstrating clearly that our most advanced technology is dependent on other advanced technology. Thus we can again ask the question: who fares best without computers, cell phones, tanks, helicopters, jet fighters, aircraft carriers, video games, and flat-screen TVs? And again, the answer is fairly straightforward: the Chinese and Indian peasants who plant crops with their own hands or domestic animals and fertilize those crops with their own feces, Kalahari Bushmen hunter-gatherers, and Australian Aborigines.

For a real doomsday scenario, one that a child born today will witness if scientific predictions about human population growth are correct, imagine the state of humanity when that population finally levels off at about nine billion. As every freshman biology student knows, the figures above actually are testable hypotheses, and a child born in 2011 will be 40-60 years old when the results of this human experiment are revealed. Today, in the United States, age 40-60 is considered the prime of life. Our consideration of questions about evolution, as outlined in the first nineteen chapters, and in the paragraphs above, leads us to a final one: what does the term "levels off" actually mean in this context; that is, when the "population finally levels off at about nine billion"?

As befitting a college prof's approach to teaching, we answer a question with a question, thus finishing our consideration of human evolution with a final exam. Please answer the following questions using knowledge you have gained so far in life by listening to the radio, watching television, reading books, surfing the Internet, going to school, going to church, talking with friends and neighbors, reading the newspapers, and using any other source of information at your disposal.

(1) Will any or all religious organizations, including the Roman Catholic Church, begin promoting birth control?

(2) Will all nations, societies, organizations, and ethnic groups agree to limit population by controlling the number of human offspring produced so that our species can live within its available resources?

(3) Will all nations, societies, organizations, and ethnic groups begin immediately to reduce, by all available means, energy and water consumption so that the "leveling off" process does not involve massive starvation?

(4) Will the governing bodies of all nations, societies, organizations, and ethnic groups begin immediately to provide free birth control information, supplies, and drugs to anyone and everyone to whatever extent is possible?

(5) Will various nations, societies, organizations, and ethnic groups continue to increase consumption of water and petroleum, while at the same time disregarding growth in human populations?

(6) Will this set of self-imposed conditions, so typical of natural systems, eventually result in a new species of "human"?

If your answers were "no," "no," "no," "no," "yes," and "yes," then you get an A.

Yes, *Homo sapiens* is evolving, and will continue to evolve, until the species becomes extinct.

20. What Will Human Life Be Like in a Couple of Thousand Years?

> *In all likelihood, any civilization that we can detect will be more advanced than our own, providing us with a glimpse of what Earth's future could be. For the first time, we will witness the history of the future, not just of the past.*
>
> —Frank Drake and Dava Sobel (*Is Anyone Out There?*)

Obviously there is no way to answer this final question for certain, but we can do a little thought experiment that might suggest some answers. Picture yourself in what is now Israel at the time of Jesus' crucifixion attempting to predict what human life would be like in the year 2012 and you will have a sense of the difficulty in making such predictions. On the other hand, what *could* you have said back then that might be applicable to the problem? The answer to this question is fairly easy: people develop technology (they build pyramids, embalm their dead, make catapults and other war machines), they practice various religions that are not always compatible, they go to war, they discriminate against other kinds of people just like we do today (gays, lepers, Philistines, etc.), they use their religion to justify behavioral dictums, including highly personal ones involving sex and diet, they practice whatever forms of agriculture are allowed by their environments, they gather food from the ocean, they use narcotics, they write, paint, and create art, they form pair bonds and produce offspring, and they engage in all sorts of political machinations.

In other words, people really have not changed much in the past two thousand years except in terms of their available technology, and we don't have much evidence that people differ very much across geographical regions, however, except in terms of cultural traits (learned), at least some of which are adaptations to environment. Pick any two humans at random from around the world, and if they are of opposite genders, then they can very likely produce viable offspring, assuming they are both of reproductive age and have not been sterilized.

In evolutionary terms, two thousand years is but a hardly measurable blink, so the answer to our question is really very simple: people two millennia from now will do exactly the same general kinds of things that people do today, except like people of today, or of Jesus' day, they will be limited by their traditions, technology, and environment. Therefore, the answer to our question depends not so much on the behavior of people themselves as on cultural traits, technological development, and environmental conditions.

Scientists and historians have given us plenty of information to use in making predictions about the general state of technological development and the environmental conditions we can expect two thousand years from now. And indeed, plenty of people have used this information and made predictions, ranging from utopian to disastrous. Regardless of this range of specific predictions by various experts, many of whom have personal or political agendas, the historical record is fairly clear. That record tells us two things: (1) you cannot predict technological innovations and developments very accurately or very far in advance, and (2) deteriorating environmental conditions are probably the most important factor in the collapse of civilizations.

Will parts of Earth experience deteriorating environmental conditions severe enough to cause collapse of various societies and consequently their civil order? The answer is "probably," not so much because I know something nobody else knows, but because the evidence is so strong that such collapses, based on environmental conditions, have happened so frequently in the past. Jared Diamond's compelling book *Collapse: How Societies Choose to Fail or Succeed* provides an easily accessible and beautifully written analysis of this relationship between society and environment and suggests that the process will be repeated again and again in the future. Diamond claims that societies collapse when their economies become unsustainable for a variety of reasons, mostly environmental. There is little evidence that Diamond is wrong and plenty that he is correct. So for the purposes of our thought experiment, let's assume that he is correct.

Of course any prediction based on this assumption rests firmly on another assumption, namely, that there will still be people on Earth in two thousand years. Two factors could invalidate this second assumption, so we probably need to deal with those factors right up front. In two thousand years, humanity easily could be extinct. We could destroy ourselves with nuclear weapons and make Earth uninhabitable, or a large asteroid could hit our planet somewhere and simply obliterate us, or at least obliterate the resources we need to survive. Invaders from outer space could arrive tomorrow and kill us all. Jesus could return to Earth tomorrow and Rapture could occur about the same time (give or take two weeks, depending on the particular version of Rapture), all believers could be immediately taken to Heaven, and the rest of the planet denuded. According to the Book of Revelation, there are a couple of thousand year periods involved in the demise of Earth, so if these prophecies are correct, then humanity as we know it will be rendered extinct by the hand of God.

Thought experiments that invoke supernatural forces are not valid ones, or even very instructive, regardless of how real the gods might be, how they might in fact behave, or what we believe about their intentions and behaviors. Indeed, the historical record is exceedingly clear on one thing: people will give the gods (including their God) credit for any natural phenomenon about which they are completely ignorant. So the historical record shows that when we learn how Nature operates, suddenly the gods disappear as causal agents although they can easily remain as political forces that inspire human behavior. But in general, people today do what people have done for all recorded history.

For this last reason, I will assume that in two thousand years, there will still be people on Earth developing technology, practicing various religions that are not always compatible, going to war, discriminating against one another, using their religion to justify behavioral dictums, including highly personal ones involving sex and diet, engaging in whatever forms of agriculture are allowed by their environments, gathering food from the ocean, using narcotics, writing, painting, and creating art, forming pair bonds and producing offspring, and engaging in all sorts of political machinations. The question of course is whether in doing all these activities they will resemble the Flintstones more than the Jetsons. Most reasonable scientists would say yes, in two thousand years humans will be living more like Fred Flintstone and his family than like George Jetson and his.

The main reason for this scientific judgment is the absence of fossil fuels, which surely will disappear during the next millennium, at least as a socio-economic phenomenon at the national and international levels. We have spent the last century converting crude oil into people, and when the crude oil is gone, then so will be the people and all those conveniences that rely on petroleum or other fossil fuels for their existence. Good examples of the latter include automobiles, motorcycles, airplanes, ships, tractors and other farm implements, warplanes, tanks, helicopters, and Humvees, many if not most electrical power generating plants, lawnmowers, gas, coal, and oil furnaces, gas stoves, most public transportation, and all manufactured goods whose production depends on fossil fuels either for energy or as chemical feedstock.

The United States already is in a state of serious competition for fossil fuels, our main competitors being India,

China, and Russia. Our nation is currently spending at least $200 million a day in a war that most reasonable people around the world believe is actually a war to control the global petroleum supply, or at least to prevent others from controlling that supply. So all the evidence indicates that we are now approaching a condition in which an exponentially increasing population has nearly used up its supply of a fixed resource. By "nearly" I mean within a few hundred years, again, but a blink in evolutionary time.

So the world of 4010 likely will be more familiar to the people of the 1st Century than to people of the 21st Century. We will use horses for transportation and our numbers will have diminished to whatever levels can be supported by agriculture without pesticides or artificial fertilizers. When we fight wars, they will be with armies that walk, or ride on horses and in wooden chariots instead of tanks, and do not fly in helicopters or jet planes. We are likely to have explosives, which the people of Roman times did not have, but the kinds and amounts are probably a matter of debate. We will fight mainly over water. The oceans will recover from their current over-fished condition, and we will probably start whaling again, once technology for building wooden ships is re-developed, a development that depends on re-establishment of adequate forests.

However, and this is a relatively important "however," we will be in this socio-economic-ecological state with our current knowledge of mathematics, art, music, and literature. It is entirely possible that the two- or three-hundred year period from the global collapse of civilization and massive death to the re-adjustment of humans to Earth-imposed limits will be one of the most miserable times in all of human history. And it is equally possible that the following thousand years will be one of the richest, culturally speaking, in all human history, mainly because our capacity for destructtion will have been so drastically reduced and we will have retained our knowledge of how to make art, music, and literature.

There is little evidence from either the historical record or the psychological research literature to support a predicttion that we will have "learned our lesson" about violence and environment destruction and achieved a purposefully harmonious relationship with our planet. Like today, many individuals will understand the need to live in such a way, but at the population level we are still going to be governed by our human genes, most of which we inherited from our non-human ancestor primates. These genes, and the behaveiors they drive us into, are described well in E. O. Wilson's classic book *On Human Nature*. Because of our genetic makeup, Wilson claims, human males will always live in dominance hierarchies populated by other males, and human females will always form mutually supportive groups capable of working together to solve problems. Our best prediction, therefore, is that two millennia hence, most if not all of us will resemble Ndani people of the New Guinea highlands more than the Silicon Valley entrepreneurs and *People Magazine* cover story subjects.

Should we worry about this transition? Probably not, because there is not much we can do about it. Even if I said yes, we *should* worry a great deal about the changes coming for humanity in the next two thousand years nobody would listen to the advice. We worry about ourselves, our next meal, our children, and if we have some and they are cute and readily accessible, our grandchildren. We don't worry much about our neighbors down the street, nor about our neighbors' great-grandchildren-to-be, even when the neighbors are good friends. In other words, our worries are about what E. O. Wilson probably would predict, based on his understanding of primate genetic heritage, namely, that we focus our emotional energies on our immediate needs and resources, those factors that threaten our small extended family troop, and various real, imagined, or metaphorical bears at the cave door.

Just for the sake of discussion, instead of trying to think two millennia ahead of today, let us envision just the next five hundred years and perhaps play another thought experiment, again by assuming that natural processes we al-ready know about will still be operational, that the fossil fuel supplies will be exhausted within the first one or two hundred years of this half-millennium, and that fresh water supplies will also become limiting (as they are close to becoming at the present time). On the technical side, to our credit, we have nuclear energy as a possible substitute for much of the fossil fuel now consumed. Because nuclear energy has the capacity to actually replace so much of the fossil fuel use, our biggest choice, as a species, is whether that nuclear energy gets used for peaceful purposes—heating and lighting homes and schools, supplying electricity for trains and streetcars, sustaining communications networks, etc.—or military one—as weapons of mass destruction both manufactured and used. So as the fossil fuel supplies begin to run out, humanity's main choice will be how to use the available technology to replace those supplies, and that technology consists mainly of nuclear power. Thus the question becomes: Are we going to use our brains and our highest, most noble character traits, exercising, in the process, unheralded rationality? Or are we going to fight.

If we decide as a species, to use nuclear weapons against one another, then the question becomes whether such use is relatively limited or catastrophic. If it is the latter, then there is no point to discussing how humans will be living in the future. If nuclear wars are limited, then the resulting elimination of infrastructure, coupled with water and alterative fuel shortages (or absences!), much of humanity will be living in urban jungles characterized by violence and low-level conventional warfare. Baghdad in 2007 is a model for such a set of conditions. In other words, in the opening years of the Third Millennium, in order to understand what happens to humans when deprived of fuel, water, services, infrastructure, and tolerance, one only needs to read the morning paper or watch Fox News. This kind of degeneration should not be particularly startling; it has happened numerous times throughout history and currently happens fairly frequently (several African nations over the past half century).

Are the conditions described above conducive to organic evolution of humans? The answer is probably, if not emphatically "yes," especially given the fact that our species is evolving quite rapidly already (see chapter 19). The next species of human is likely to be darker, on the average, than the present one, and will be smaller and smarter in the same way that certain so-called primitive tribes are smart. In other words, the environment will dictate survival much more than we believe it does today, and those individuals who are mentally adept at picking up survival skills will produce the most offspring. What are those skills likely to be?

Again, the question is fairly easy to answer because we have so many good examples from the past century. These people will be highly resourceful, capable of forming small group alliances when necessary, and their digestive system bacterial flora will be quite different from ours because of the mixture of foods they will be forced to eat. Those foods will be about anything that contains protein, and in the post-petroleum world the most readily available source of protein will be rats, cockroaches and other insects, chickens, pigeons, cattle, horses, dogs, and cats. For people living near the coasts, molluscs will become a major part of the diet, just as they were millennia in the past. The potential food species are generalists who seem to survive quite well under stressful conditions, and aside from the cats, are reasonably omnivorous. Vegetable components of the future humans' diet will be about anything that grows and is not poisonous. Finally, the population density of Fifth Millennium people will be much lower than at present, and large cities, if they exist, will be quite isolated from one another physically but probably not in terms of communication.

If there are large animals present two millennia hence, those are most likely to be horses, camels, or similar types of livestock, depending on the region. In other words, people will keep animals that serve multiple functions: food, transportation, and labor. Coastal regions are likely to flourish because the capacity for destruction of marine life will be so drastically reduced. Thus whaling is very likely to make a comeback. This comeback will depend on forests, as mentioned above, to provide wood for ships, and on the ability of oceans to support a food chain leading up to very large mammals. That ability depends in large part on radiation, algae, and pelagic crustaceans; provided we have not completely destroyed the ozone layer to the point of bathing the planet in ultraviolet rays, then the oceanic algal populations are likely to flourish, which means that the remainder of marine food chains are also likely to flourish. So among the most fortunate and stable populations of the Fifth Millennium will be Eskimos, provided they retain their ancestral knowledge of how to build kayaks and other whaling craft from available materials. The whaling industry in more temperate climes could easily rebound to what it was in the 18th and 19th Centuries in the United States, but the industry is likely to experience cycles resulting from unregulated harvest. This time around, the meat will be consumed as well as the oil.

Will there be wars? Most certainly there will be wars, although it's a little difficult to predict what the weapons are likely to be, beyond those that were present prior to the discovery of petroleum. So as a minimum, we should expect warfare similar to that of the Romans. However, there is one human trait that we should not ignore in this analysis of the future, and that is our ability to retain knowledge gained over the past. Humans of the Fifth Millennium will be exceedingly cunning in warfare, and perhaps exceedingly cruel, too. We can expect extremes of torture and mutilation, coupled with social insults that produce major emotional trauma at the individual as well as social levels.

Good examples of this latter situation already exist. Beheading of captives, especially when done slowly and in a ritual way, is emotionally wrenching, and in the absence of the Internet, we might well expect creative solutions to the communication problem. Many such solutions come immediately to mind, but I'm reluctant to mention them in case someone might take them to heart now instead of waiting another two thousand years. And as for social trauma, elicited by words and pictures, one need only remember the Islamic reaction to those Danish cartoons depicting The Prophet Muhammed in less than flattering situations. There is little reason to believe that such conflict can be resolved through political means, and there is every reason to believe that in the future, wars will continue unabated for decades, even becoming ritualized, as they did in Ndani societies of New Guinea.

So the major question regarding humanity's future is whether we will retain our electronic communication skills beyond the Age of Fossil Fuels. In the absence of truly catastrophic nuclear Armageddon humanity should be able to retain much of its electronic communication technology. If we subtract fossil fuels from the communications industry as it now exists then we lose repair vehicles, much of the manufacturing capacity, and energy supply. The latter can be replaced easily by nuclear power, but manufacturing and maintenance components remain a major question mark. Indeed, the future of electronic communications may well be the biggest question facing humanity over the next two thousand years. If we lose it, then our future is reasonably predictable although not altogether pleasant (see above); if we retain it, then our future is very unpredictable. Having already addressed the former case, let's consider some possibilities for the latter.

By assuming that we retain electronic communication capacities, we are also assuming that we somehow manage to manufacture, maintain, and distribute communication devices and maintain electronic networks. Given these assumptions, then, the question becomes: What kind of a human society might develop with rapid global communications but without rapid global travel? We have a wide spectrum of choices. At the positive end of this spectrum is an almost utopian world in which people share ideas, literature, images, and advice on living within available resources; at the negative end is a world of extreme cunning in which our primitive traits are enhanced, strengthened, and manipulated to an extreme as groups compete with one another.

In the latter case, males will operate in dominance hierarchies characterized by skill at use of information, females will be quite subordinate and relegated to groups integrated and controlled by information, and wars will be fought slowly but with extreme imagination and innovation regarding the use of available weapons. This latter condition will put a premium on male intelligence, attention span, discipline, and creativity, something that is generally lacking among a large section of today's male population, at least in the United States. Thus we can anticipate a warrior class, a labor class, and an oligarchy of really smart and cunning males.

Given the record of *Homo sapiens* on Earth to date, I suspect that the utopian alternative is little more than wishful thinking and that war will be the norm, as it generally has been for much if not all of recorded history. Even today, in this grand experiment we call the United States of America our circumstances and official actions are beginning to resemble those spelled out in George Orwell's *1984*. The world is divided roughly into three main economic, religious, and social units: Asia, the Islamic Middle East, and North America-Europe. In the United States we have an Office of Homeland Security, the Patriot Act, untold thousands of video monitors in places ranging from convenience stores to busy intersections and a hyperactive religious community with clear beliefs regarding Armageddon, Rapture, and the return of Jesus.

We also have a diverse set of electronic gadgets that take information and send it around to all kinds of places at the speed of light: IP address checkers, Radio Frequency Identification Tags in credit cards, clothing, and pets, as well as similar identification chips embedded beneath our skin (if we want one), caller ID, spyware that reports your online browsing habits, and online shopping sites that survey your buying habits and start recommending items for purchase. Winston Smith, George Orwell's Ministry of Truth bureaucrat, would see a lot of familiar items and circumstances.

Humans are, of course, quite different from other animals in the extent to which we are self-aware, thus cognizant of our past and curious about our future. But humans also require food, water, and shelter, thus are certainly no different from all other organisms in this respect. Water may exist as a fluid, solid, or gas, but it's always H_2O. Our food comes many various forms, but it's ultimately derived from plants, fungi, animals, and microbes—that is, non-human organisms, the ones with which we share the planet. In terms of potential food resources for the future, there is one prediction that is absolutely accurate: we are living in an age of mass extinction, and we are the cause.

Destruction of tropical forests alone validates this prediction because scientists estimate that about seventy percent of all genetic information present on Earth resides in these forests. That genetic information comes primarily in the form of microbes, plants, and animals, especially insects. Although most of these species, especially the plants, probably are inedible, many also may play important roles in pollination and seed dispersal, processes that are essential for the health of terrestrial ecosystems. We are now obliterating this wealth of biological resources at a frantic pace, producing not only agricultural products, at least in the short term, but mass extinction, social upheaval, political conflict, and additional cycles of environmental destruction and its consequences. Humans are simply killing off the world's biota faster, and more effectively, than it has been destroyed by natural causes at any time in Earth's history.

The fossil record tells us a great deal about mass extinctions because it provides evidence that several have occurred over the past half-billion years and allows us to track the long-term consequences of such extinctions. At the end of the Permian Period, approximately 230 million years ago, at least 90% of all known genera disappeared. The disappearance of North American megafauna (mammoths, ground sloths, etc.) was a much more recent event that some have attributed to human activity, at least in part. The net effect of mass extinctions is substantial loss of biological diversity, followed by evolutionary diversification from the remaining genetic stock.

Animal life today, for example, is far less diverse, in terms of basic body architecture, than it was 500 million years ago. Stephen J. Gould documents this reduction of diversity in his book *Wonderful Life: The Burgess Shale and the Nature of History,* showing us what Earth has lost in terms of exotic creatures. Such loss of genetic diversity, followed by subsequent diversification from a much-reduced supply of genes, is called a *genetic bottleneck*. Cheetahs and pandas are well-known cases that illustrate the idea. *Homo sapiens*, with its enormous capacity for environmental destruction, may well be the champion of all evolutionary forces, generating bottlenecks more rapidly than, and perhaps surpassing, those produced by continental drift, climate change, and volcanic catastrophe over the past billion years.

Two thousand years equals about a hundred human reproductive cycles, certainly enough time for our species to become noticeably smaller and darker, especially if confronted with no petroleum, rapidly increasing global temperature, and limited freshwater supplies. If we extend our thought experiment to two million rather than two thousand years and ask what the descendents of *Homo sapiens* will be like, then the potential answers become very interesting indeed. Much of this interest stems from the fact that we already have a model for primate evolution over a similar period in the past: the divergence of chimpanzees from early species of the genus *Homo*.

We also have a model for major changes in large animals over extended periods of time following mass extinctions: the evolution of dinosaurs into modern birds. For a truly imaginative scientist with these models, it's not too difficult to conjure up an extremely intelligent and cunning little brown primate, perhaps the size of a squirrel, eating rodents and insects, cultivating certain plants, using poison darts for defense against predators, sitting around tiny fires singing magically beautiful quiet songs and telling novels to adolescent children the size of today's mice.

References and Sources:
Books:
Beard, C. 2004. *The Hunt for the Dawn Monkey*. University of California Press, Berkeley, CA.

Behe, M. 1996. *Darwin's Black Box: The Biochemical Challenge to Evolution*. The Free Press, New York, NY. 307p.

Coleman, S. and L. Carlin (eds) 2004. *The Cultures of Creationism: Anti-evolutionism in English-Speaking Countries*. Ashgate Publishing, Ltd., Burlington, VT 195p.

Darwin, C. 1859. *On the Origin of Species by Means of Natural Selection, or the Preservation of Favoured Races in the Struggle for Life*. John Murray, London. 502p.

Dawkins, R. 2006. *The God Delusion*. The Bantam Press, New York, NY.

Dembski, W. A. 2002. *No Free Lunch: Why Specified Complexity Cannot be Purchased Without Intelligence*. Rowman and Littlefield, Lanham, MD. 404p.

Dembski, W. A. (ed) 2004. *Uncommon Dissent: Intellectuals Who Find Darwinism Unconvincing*. ISI Books, Wilmington, Delaware 366p.

Dyson, F. 1984. *Weapons and Hope*. Harper and Row, New York, 340 p.

Frank, T. 2004. *What's the Matter with Kansas? How Conservatives Won the Heart of America*. Metropolitan Books, New York. 306p.

Fussell, P. 1989. *Wartime: Understanding and Behavior in Second World War*. New York: Oxford University Press, 330p.

Godfrey, L. R. 1983. *Scientists Confront Creationism*. W. W. Norton & Co., New York, 324 p.

Gunn, A. M. 2004. *Evolution and Creation in the Public Schools.* McFarland & Company, Inc., Jefferson, NC 187p.

Harrold, F. B., and R. A. Eve (eds) 1987. *Cult Archeology and Creationism: Understanding Pseudoscientific Beliefs About the Past.* University of Iowa Press, Iowa City, IA.

Hayward, J. L. 1998. *The Creation/Evolution Controversy: An Annotated Bibliography.* Scarecrow Press, Lanham, Maryland; Salem Press, Pasadena, California. 253p.

Isaak, M. 2007. *The Counter-Creationism Handbook.* University of California Press, Berkeley. 330p.

Kowalewski, M. R. 1990. Religious constructions of the AIDS crisis. *Sociological Analysis* 51:91-96.

Lindberg, D. C. and R. L. Numbers, eds. 1986. *God and Nature: Historical Essays on the Encounter between Christianity and Science.* University of California Press, Berkeley. 516p.

Locke, S. 1999. *Constructing "The Beginning": Discourses of Creation Science.* Lawrence Erlbaum and Associates; Mahwah, NJ 235p.

McIver, T. 1988. *Anti-evolution: An Annotated Bibliography.* McFarland, Jefferson, NC

Montagu, A., and F. Matson. 1983. *The dehumanization of man.* New York: McGraw-Hill Publishing Company, 246p.

Moreland, J. P. (ed) 1994. *The Creation Hypothesis: Scientific Evidence for an Intelligent Designer.* Inter-Varsity Press, Downers Grove, IL 335p.

Pennock, R. T. (ed) 2001. *Intelligent Design Creationism and its Critics: Philosophical, Theological, and Scientific Perspectives.* MIT Press, Cambridge, MA 805 p.

Pigliucci, M. 2002. *Denying Evolution: Creationism, Scientism, and the Nature of Science.* Sinauer Associates; Sunderland, MA

Smout, K. D. 1998. *The Creation/Evolution Controversy: A Battle for Cultural Power.* Praeger Publishers, Westport, CT 209p.

Thomas, R. M. *Religion in Schools: Controversies Around the World.* Praeger Publishers, Westport, CT

Tourney, C. P. 1994. *God's Own Scientists: Creationists in a Secular World.* Rutgers University Press, New Brunswick, NJ 289p

Tuchman, B. 1984. *The March of Folly: From Troy to Vietnam.* New York, Alfred A. Knopf.

Wilson, D. B. (ed) 1983. *Did the Devil Make Darwin Do It?: Modern Perspectives on the Creation-Evolution Controversy.* Iowa State University Press, Ames, IA.

Web sites (active in 2011), newspaper stories, and other sources:

It's a well-known fact that web sites, newspaper stories, etc., vary widely in their accuracy as well as their polemical content. The ones listed below certainly illustrate this fact. For many of these topics, one could easily list many more sites, so it's quite possible that your favorite ones are missing from the lists below.

Chapter 1 – What is Intelligence?

New York Times, Wednesday, July 18, 2007

http://www.smh.com.au/news/world/hurricane-is-gods-work-christian-extremists/2005/...

http://www.smh.com.au/news/world/hurricane-is-gods-work-christian-extremists/2005/09/03/1125302770141.html

http://www.avert.org/needle-exchange.htm

http://www.statehealthfacts.org/comparetable.jsp?cat=11&ind=566

Chapter 2 – What is Design?

http://dot5.drawar.com/posts/what-is-design

http://atschool.eduweb.co.uk/trinity/watdes.html
http://experiencedynamics.blogs.com/site_search_usability/2007/10/what-is-design-.html
http://en.wikipedia.org/wiki/Design

Chapter 3 – What is Intelligent Design?

http://ncse.com/creationism/general/what-is-intelligent-design-creationism?gclid=CL-a25Pw4agCFapl7AodLhR3HA
http://www.intelligentdesign.org/whatisid.php
http://www.arn.org/idfaq/What%20is%20intelligent%20design.htm
http://en.wikipedia.org/wiki/Intelligent_design
http://www.discovery.org/csc/topQuestions.php

Chapter 4 – What is Complexity?

http://pespmc1.vub.ac.be/complexi.html
http://en.wikipedia.org/wiki/Complexity
http://evolution-101.blogspot.com/2006/05/what-is-irreducible-complexity.html
http://www.complexitynet.eu/science/Pages/default.aspx
http://nirmukta.com/2009/08/18/complexity-explained-1-what-is-complexity/

Chapter 5 – What is Creationism?

http://www.infidels.org/library/modern/science/creationism/
http://ancienthistory.about.com/od/creationmyths/a/CreationMyths.htm
http://www.theologicalstudies.org/classicalreligionlist.html
http://www.answersingenesis.org/creation/v14/i1/fossil.asp
http://answers.yahoo.com/question/index?qid=20081024111307AAq0ger

Chapter 6 – What is Science?

http://www.gly.uga.edu/railsback/railsback_1122science1.htm
http://undsci.berkeley.edu/article/whatisscience_01

http://www.aps.org/policy/statements/99_6.cfm
http://www.pbs.org/wgbh/evolution/library/09/index.html

Chapter 7 – What is Religion?

http://www.teachingaboutreligion.org/WhitePapers/Larue_whatisreligion.htm

http://atheism.about.com/od/religiondefinition/a/definition.htm

http://en.wikipedia.org/wiki/Religion

Chapter 8 – What is a Conflict Between Science and Religion?

http://en.wikipedia.org/wiki/Large_Hadron_Collider

http://www.cbsnews.com/stories/2004/03/15/national/main606202.shtml

http://pmo.vox.com/library/post/do-you-believe-in-evolution.html (do you believe in evolution – asked of Huckabee)

Chapter 9 – Why are Science and Religion in Conflict?

http://www.iep.utm.edu/sci-rel/

http://etext.virginia.edu/toc/modeng/public/DraHist.html

http://mwillett.org/atheism/relsci.htm

http://www.psk12.com/rating/index.php

Chapter 10 – What is Evolution?

http://en.wikipedia.org/wiki/O._J._Simpson_murder_case

http://www.talkorigins.org

Chapter 11 – What Kinds of Organisms Share the Planet with Us?

http://en.wikipedia.org/wiki/Shingebis

Chapter 12 – What is Taught in Biology Class?

http://www.britannica.com/bps/additionalcontent/18/33304224/What-Is-Taught-in-Biology-Why-Does-it-Matter

http://www.hampshire.edu/news/16376.htm

Chapter 13 – What Should be Taught in Biology Class?

http://www.nature.com/nrm/journal/v7/n4/abs/nrm1856.html

http://www.ncbi.nlm.nih.gov/pmc/articles/PMC2104500/

Chapter 14 – What is the Meaning of "Science Literacy"?

http://en.wikipedia.org/wiki/Levonorgestrel

http://www.arhp.org/healthcareproviders/onlinepublications/clinicalproceedings/lngiusguide/levonorgestrel_brochure.cfm?ID=96

http://www.austinchronicle.com/gyrobase/Issue/story?oid=oid%3A394228

http://www.womensenews.org/article.cfm/dyn/aid/2526/context/archive

http://www.thenation.com/doc/20050718/mcgarvey

http://content.nejm.org/cgi/content/short/350/15/1561

Chapter 15 – Why is Science Literacy so Important?

http://stats.org/stories/2008/needle_exchange_drug_czar_dec03_08.html

Chapter 16 – Why are Politicians so Scientifically Illiterate?

http://www.energytribune.com/articles.cfm?aid=2210

http://books.google.com/books?id=oaqp7byW99YC&pg=PA183&lpg=PA183&dq=why+are+politicians+so+scientifically+illiterate&source=bl&ots=VpXi1pyCiQ&sig=J7E-rva1rJnH2alviksz6Lfb_rI&hl=en&ei=3aJTTv-DJomJsgLO1KXBBw&sa=X&oi=book_result&ct=result&resnum=7&ved=0CDwQ6AEwBg#v=onepage&q=why%20are%20politicians%20so%20scientifically%20illiterate&f=false

http://www.goodreads.com/book/show/5741758-unscientific-america

Chapter 17 – Is "Evolution" Dangerous?

http://en.wikipedia.org/wiki/The_Bible_and_history

http://en.wikipedia.org/wiki/Mephisto_Waltzes

Chapter 18 – What is a Human Being?

http://www.aboutseafood.com/media/facts_statistics_detail~id~0.cfv

Chapter 19 – Are Humans Evolving?
http://www.pfaf.org/user/edibleuses.aspx
http://www.fao.org/WAICENT/faoINFO/AGRICULT/AGL/aglw/cropwater/wheat.stm#supply
http://www.nue.okstate.edu/Crop_Information/World_Wheat_Production.htm
http://www.ehow.com/list_6162960_materials-used-make-computers_.html
http://www.ehow.com/list_6757391_materials-used-make-cell-phones_.html

Chapter 20 – What Will Human Life be like in a Couple of Thousand Years?
http://www.wired.com/wired/archive/8.04/joy.html

The Author:

John Janovy, Jr. has published eighteen books, fifteen of which are trade publications; most of these books use nature to explore themes such as interdependency, intellectual freedom, testing one's limits and learning from the experience, creativity in teaching, and the development of careers, but some are fiction, including science fiction. Hs also is the co-author (with Larry Roberts) of *Foundations of Parasitology*, the world's leading text in a general subject area that includes tropical disease economics and transmission, exceedingly complex epidemiology, enormous biological diversity, and infectious agents with which humans interact daily. Dr. Janovy also wrote the script for the PBS film version of *Keith County Journal*, which was televised internationally in the late 1980s.

In addition to the books, he has published nearly 100 scientific papers and advised a long list of now-successful graduate students and honors undergrads. Dr. Janovy has taught, often to very large audiences, for 45 years and has won numerous teaching awards, including the University of Nebraska Distinguished Teaching Award, Burlington-Northern Teacher-Scholar Award, University of Nebraska Outstanding Research and Creativity Award, and the American Society of Parasitologists Clark P. Read Mentor Award.

Dr. Janovy lives and works in Lincoln, Nebraska. He and his wife Karen have three grown children. Details of his professional life, including presentation audios. PowerPoints and a complete CV, can be found at:

http://www.johnjanovy.com.

His blog is http://talkparasites.blogspot.com.

www.ingramcontent.com/pod-product-compliance
Lightning Source LLC
Chambersburg PA
CBHW031830170526
45157CB00001B/247